THE M. & E. HANDBOOK SERIES

BRITISH AND AMERICAN ECONOMIC HISTORY
1850–1950

THE M. & E. HANDBOOK SERIES

BRITISH AND AMERICAN ECONOMIC HISTORY 1850–1950

R. WINDLE, B.SC.(ECON.)

Principal Lecturer and Course Supervisor, Department of Social Studies, Portsmouth Polytechnic

MACDONALD & EVANS LTD.
8 John Street, London, W.C.1.
1971

First published February 1971

©

MACDONALD AND EVANS LIMITED
1971

S.B.N.: 7121 0225 6

*Printed in Great Britain by The Garden City Press Limited,
Letchworth, Hertfordshire*

AUTHOR'S PREFACE

THIS **HANDBOOK** is intended to serve as a general introduction to the economic history of Britain and the United States of America, and to provide information in a convenient form for those preparing for examinations, such as first-year degree students reading British and American history and candidates preparing for the General Certificate of Education Advanced Level examinations or the Scottish Certificate of Education Economic History on the Higher Grade examination. In addition the sections on Britain should meet some of the needs of candidates at the Intermediate Social and Economic History examination of the Local Government Examinations Board.

Students should work through the book systematically and attempt the Progress Tests as chapters are read. It is important to refer to both the standard and the specialist works listed in the Bibliography, which includes the titles of learned articles from the *Economic History Review* (published for the Economic History Society), the *Journal of Economic History* (an American publication), the *Economic Journal* (published for the Royal Economic Society) and *Oxford Economic Papers* (an Oxford University journal). Additional articles can be found in other periodicals such as *Economica*, *Past and Present* and the *Agricultural History Review*. The student is urged to read the latest articles in these periodicals and to look at book reviews so that opinions and conclusions can be kept up to date in the light of the most recent research. Relevant statistics are to be found in:

B. R. Mitchell and P. Deane: *Abstract of British Historical Statistics* (Cambridge University Press, 1962);

P. Deane and W. A. Cole: *British Economic Growth, 1688–1959*, second edition (Cambridge University Press, 1967);

The British Economy: Key Statistics, 1900–1966 (Times Newspapers Ltd.);

Historical Statistics of the United States: Colonial Times to 1957 (United States Government Printing Office, 1960).

The American Historical Association's Service Center for Teachers of History publishes a number of pamphlets of relevance to the economic history of the period covered by this **HANDBOOK**.

I am indebted to the Senate of the University of London for permission to reproduce questions from B.Sc. (Economics) examination papers and to the Scottish Certificate of Education Examination Board for permission to reproduce questions from the Economic History on the Higher Grade examination papers. I am grateful, too, to Cambridge University Press for permission to make use of statistical tables in:

B. R. Mitchell and P. Deane: *Abstract of British Historical Statistics*;

P. Deane and W. A. Cole: *British Economic Growth, 1688–1959* (second edition);

Sir J. H. Clapham: *An Economic History of Modern Britain.*

In conclusion it should be noted that where the term "billion" is used in this book its meaning differs according to whether it is used in connection with Britain or the U.S.A.; in Britain it means a million millions, but in the U.S.A. it means a thousand millions.

November 1970 R. W.

NOTICE TO LECTURERS

Many lecturers are now using **HANDBOOKS** as working texts to save time otherwise wasted by students in protracted note-taking. The purpose of the series is to meet practical teaching requirements as far as possible, and lecturers are cordially invited to forward comments or criticisms to the publishers for consideration.

P. W. D. REDMOND
General Editor

CONTENTS

vii

PART FOUR: INDUSTRIAL DEVELOPMENT

PART SEVEN: FOREIGN TRADE

PART EIGHT: BANKING

CONTENTS   xi

PART ONE

INTRODUCTION

CHAPTER I

PHYSIOGRAPHY, GOVERNMENT, WAR AND CYCLICAL FLUCTUATIONS

PHYSIOGRAPHY

1. Area and climate.

(*a*) The area of the U.S.A. (without including Alaska and Hawaii) exceeds 3 million square miles (7·7 m. sq. km): the area of Britain (England, Wales and Scotland) is less than 97,000 square miles (250,000 sq. km).

(*b*) The U.S.A. extends north to south over more than 20 degrees of latitude and east to west across more than 55 degrees of longitude, whilst Britain is encompassed within 9 degrees of latitude and 8 degrees of longitude. Consequently, the U.S.A. has greater climatic variations than has Britain, where no area is more than approximately 70 miles (112 km) from the sea and where the warm Gulf Stream helps to even out seasonal variations in temperature.

(*c*) Rainfall is one determinant of crop production. The whole of Britain receives adequate rainfall for agriculture of differing types, but the east is drier than the west.

In the U.S.A. land east of the hundredth meridian is relatively sure of adequate rainfall for agricultural purposes, but west of this, except near the west coast, the average rainfall is less than adequate. The main agricultural areas are, therefore, east of the hundredth meridian and near the west coast.

(*d*) Climate and position have influence over agriculture. A greater variety of crops can be produced in the U.S.A., on a commercial scale, than in Britain. The size of the area suitable for food production in the U.S.A. per head of population is much larger than Britain's.

2. Relief and internal trade. The river systems of both countries

1

provide access to the interior, but mountain ranges have presented more obstacles to communications and transport in the U.S.A. than in Britain. Prior to 1850 the flow of trade in the interior of the U.S.A. was determined by the Appalachian and Adirondack mountain ranges and by the Mississippi–Missouri–Ohio river systems: trade was pulled in a north-to-south direction. The Rocky Mountains presented obstacles to the westwards extension of the frontier and to the construction of transcontinental rail-roads. (Students should familiarise themselves with the physical features of the U.S.A. and Britain, and should know the location of the fifty states that constitute the U.S.A.)

The opening of the Erie Canal, linking New York with the Great Lakes, began to change the flow of trade from north to south to west to east. Further canal construction and then railway construction emphasised the west-to-east trade flow. In Britain canals changed trade-flow patterns to some extent, but railways largely emphasised the build-up of existing patterns, with London as the magnet.

3. Natural resources.

(a) Indigenous raw materials have been of greater economic significance in the U.S.A. than in Britain in the nineteenth and twentieth centuries. The former country became a leading exporter of raw materials and of foodstuffs before the First World War, exporting vast quantities of raw cotton, naval stores, tobacco, wheat, corn and meat. Coal was the only indigenous raw material that Britain exported in large quantities.

(b) Britain was and still is a food and raw material deficit area. She relied upon increasing her exports of manufactures and increasing invisible exports to enable her to pay rapidly increasing import bills for foodstuffs and raw materials. The U.S.A. was and remains a food surplus area, but indigenous supplies of some raw materials were used up rapidly between 1865 and 1920. For example, timber was imported in significant quantities by 1950 although huge reserves existed in 1850. Raw materials became relatively less significant in the total of American exports as industry expanded.

GOVERNMENT AND ECONOMIC AFFAIRS

4. The Houses of Parliament and Congress.

(a) In Britain government is effected by *Parliament* – the

House of Commons, the House of Lords and the Monarch. The executive powers are in the hands of the Prime Minister and his colleagues in the Cabinet. In the U.S.A. *Congress* is the ruling federal body, consisting of the House of Representatives and the Senate. Executive powers are in the hands of the President and his Cabinet.

(b) Britain is *a unitary state*, whilst the U.S.A. is *a federal nation* in which laws are both federal and state in origin. State intervention in the U.S.A., therefore, can refer to intervention either by the federal authority or by the state governments.

5. State intervention in economic affairs. In both countries governments have intervened in economic affairs, either to give impetus to economic development or to control aspects of it. State governments in the U.S.A. aided transport and banking developments, in particular, by direct investment or by using state credit-worthiness to assist in the raising of funds. The federal government provided aid to initiate development (*e.g.* providing railway companies with large areas of free land to promote settlement and railway construction).

(a) State intervention in *Britain* until recent times was not concerned so much with initiating economic development but rather with directing it along certain lines and curbing any undesirable effects that resulted from it. Railway companies, canal companies, shipping companies, etc., in the nineteenth century did not have recourse to direct state aid (apart from mail contracts). As private enterprise gave rise to problems for the community as a whole, intervention was induced to control undesirable aspects of development. Conditions and hours of work were controlled; public-health authorities were set up to control sanitation; wage bargaining was subject to legislation; state schemes for financial provision for those suffering from sickness, for the unemployed and for the aged were introduced; state educational facilities were provided and extended. In the inter-war period the government had to come to the aid of industry by changing the commercial policy to one of protection, instituting marketing cartel-type schemes, providing assistance for distressed areas and promoting a rationalisation of industry. Since 1945 much of industry has been brought under state control; the economy is a mixture of free enterprise, state control and state planning.

(*b*) State intervention in *the U.S.A.*, apart from direct and indirect aid to stimulate economic growth, occurred, as in Britain, to curb and control some aspects of development. In the late nineteenth century largely unsuccessful attempts were made to curb and control large-scale business units by means of anti-trust legislation. Industry was safeguarded in the domestic market by protective tariffs. Since 1929 there has been an accelerated growth of state intervention.

6. The New Deal. In 1929 the American economy, after experiencing from 1922 extremely prosperous times, in which the era of "high mass consumption" dawned, entered a state of chronic depression. So fraught with economic difficulties were the years 1929–32 that Roosevelt had to institute his New Deal policies in an attempt to prime the economic pump of the nation. State intervention was considered necessary in banking (the *Emergency Banking Act*), in the financial affairs of the stock exchange (the *Securities Exchange Act*), in industry generally (the *National Industrial Recovery Act*), in agriculture (the *Agricultural Adjustment Acts*), in the electric power industry (the *Muscle-Shoals – Tennessee Valley Development Act*), in transport (the *Railway Emergency Act*, the *Motor Carrier Act*, the *Air Mail Act* and the *Merchant Marine Act*), in relief and security of workers (the *Emergency Relief Act* and the *National Employment Service Act*), in safeguarding the labour movement (the *National Labour Relations Act*) and in foreign trade (the *Reciprocal Trade Agreements Act*).

(*a*) *The New Deal was of great economic significance.* It was the first time the federal government had brought to bear its powers to pull the economy out of a depression by direct intervention, and it implied a much greater participation in economic life by the government than ever before. Just as British governments gradually abandoned *laissez-faire*, so the U.S.A. gradually abandoned it in the nineteenth and twentieth centuries. The New Deal amounted to a massive acceleration of the process.

(*b*) *The success of the New Deal is not easy to measure.* Its aims were as follows:

(*i*) To increase demand by ensuring labour had a fair share of the national income.

(*ii*) To maintain prices at a reasonably profitable level by controlling marketing and output – which implied acknowledgment of the role of the large combinations of firms (this aim conflicted with (*i*) above).

(*iii*) To prime the pump by deficit financing and by the use of public works projects.

(*iv*) To reactivate foreign trade and to lower tariffs by agreements with other nations.

(*v*) To stabilise banking.

Industrial output gradually recovered, foreign trade began to revive and banking became more stable, but unemployment remained high and in 1938 another recession set in and was halted only by the onset of the Second World War. Agriculture, coal-mining, shipbuilding, etc., continued to be in difficulties. The New Deal provided some economic stimulation for recovery, but the priming of the pump was not sufficient because federal expansionist policy was nullified by the conservative policies of the various state governments.

7. Economic growth and economic planning. In the early nineteenth century it was considered that economic growth would follow automatically from the free working of an economy in which individuals pursued their own self-interests, and substantial economic growth resulted from the uncontrolled activities of entrepreneurs in both the U.S.A. and Britain. In some cases great social costs were involved for society by such activities, *e.g.* soil erosion in parts of the U.S.A., river pollution, insanitary tenements in urban areas (which facilitated the spread of disease) and so on.

(*a*) Governments in both countries were drawn into the economic sphere in two ways:

(*i*) It was necessary to control certain aspects of development in the interests of society as a whole and this could be done only through government intervention and the imposition of laws by the legislature.

(*ii*) The financial commitments of British and American governments have increased enormously. In Britain the social services and defence involve the government in massive economic activity: in the U.S.A. military budgets have a great influence on the economy. Tax changes and legislation impinge on economic activity: competition is not free in either country; the state intervenes directly and indirectly.

(*b*) Economic planning by governments has become an important factor in economic growth. Additionally, the rise of the giant business firm entails small groups of businessmen controlling enormous budgets in the private sector, too. Technology changes fairly rapidly and to take full advantage of changes planning is necessary within the firm and in the state. Whereas intuitive rule-of-thumb methods might have sufficed in 1850, by 1950 great financial and technical expertise had become essential to ensure that sustained economic growth continued on a relatively smooth path. Economic philosophy changed between 1850 and 1950; by the latter year policies of full employment were consciously pursued by governments, and this involved considerable economic planning.

WAR AND ECONOMIC GROWTH

8. Economic causes of the American Civil War. Controversy exists among historians as to the relative importance of the various causes of the war, but the main causes were as follows:

(a) *Slavery.* Most writers come back to slavery as the root cause of the Civil War. Some assert that the southern slave economy was in difficulty because of rising costs and that an expansion into new areas was necessary if slavery was to remain profitable. However, more recent research of an econometric nature has led to this view being disputed. It is argued that slavery was profitable and would not have collapsed of its own accord, although it is admitted that an expansion of slave territory would have made it more profitable. Vested interests supported the retention of slavery and explain southern interest in secession.

(b) *Settlement of the West.* The prospect of an expansion of slave territory was unacceptable to many in the North but desired by planter interests in the South.

(c) *Railroad policy.* Northern and southern interests conflicted over transcontinental railroad plans.

(c) *Tariffs.* Whilst the tariff was a bone of contention, the issue was not a simple one of North versus South: northern industrialists differed about the need for a high tariff barrier.

(d) *Banking and credit facilities.* Again the controversy about banking and credit was not a simple division between the North and South; it was linked with friction between debtor areas and creditor areas.

(e) *Immigration and the supply of labour.* Attempts to pass the *Homestead Act* were thwarted by the South until after secession, because the Act was seen as an inducement to a further growth of immigration accompanied by an increased westward movement of small farmers that would increase the number of areas hostile to slavery. However, controversy about immigration and the labour supply was not solely a North-versus-South argument. Vested interests in both North and South wanted a limitation of immigration, but the key issue was whether the labour force should be expanded as a "free" labour force or a slave labour force.

Summary. The causes of the Civil War do not fall into simple categories, since clashes of opinion about the tariff, banking, railroads and the labour supply were not simple North-versus-South controversies. Slavery seems to have been the basic issue around which southern secessionists coalesced: the elimination of slavery required federal legislation and, as events transpired, this could be effected only after a military conflict in which the North emerged victorious.

9. The economic effects of the American Civil War. These can be said to be as follows:

(*a*) *Short-term effects:*

(*i*) Production of textiles, iron and steel, machinery and the construction of railways increased during the war years, and there was a gathering pace of industrialisation from 1865 onwards, especially in the North. However, large economic losses were sustained, especially in the South. The southern export trade almost ceased because of the naval blockade; this had dire effects on the cotton planters and on Lancashire textile producers, who suffered from a shortage of raw material. Heavy destruction of property occurred in the South.

(*ii*) The total cost is estimated to have been about $9 billion, with human casualties of about 1 million.

(*iii*) It is arguable that industrial growth was retarded by the war, and that the rate of growth of the metallurgical industries, the textile industries and railroad construction would have been greater had no armed conflict occurred.

(*b*) *Long-term effects:*

(*i*) Slavery was ended, as was the southern domination of Congress.

(*ii*) Sharecropping became common in the South. Protective tariffs were retained – not because of consensus among industrialists in the North, but because of the government's need of revenue.

(*iii*) Railway construction increased – aided by the elimination of southern interests in Congress in the war. Transcontinental railroad plans were not obstructed.

(*iv*) It was considered that the industrial revolution was speeded up because of the predominance of business interests in the party in power – the Republicans. Probably industrial growth, when one compares 1865–75 with 1850–60, was much greater, possibly resulting from the effect of inflation during the war transferring income from wage and salary earners to profit makers who were willing to invest and thereby stimulate further growth. Measurement of this is not possible: estimation must suffice.

(*v*) Investment in the South was retarded: problems of reconstruction lingered on for a generation or so, and it is estimated that capital investment in southern manufacturing as a proportion of the national to total investment had diminished to about 10 per cent by 1880.

Apart from settling the issues of slavery and secession, the Civil War does not appear to have created a "great divide" in the growth of the economy: in the North it was marked by a slowing down in industrial growth (but growth continued); in the South it meant actual economic loss, but recovery was relatively rapid although economic growth was not so great as in the North in the decade after the war.

10. The impact of the two world wars.

(*a*) The roles of Britain and the U.S.A. in the international economy were affected. Britain, the major creditor nation in 1914, had to sacrifice many of her overseas investments in both wars. Even by 1918 the U.S.A. had become the major creditor on current account: after 1945 she was the lynchpin of the international economy. On her willingness to pour in aid in many forms depended the economic recovery of Europe and much of the sterling area. Whilst the U.S.A. emerged from both wars in a strong economic position, Britain was in an unhealthy economic condition by the end of the Second World War, with acute balance of payments difficulties. International roles were almost reversed between 1913 and 1945.

(*b*) The First World War upset the established system of multilateral trade based on the gold standard. Nations resorted to protection in the inter-war period, led by the U.S.A., but after the Second World War the U.S.A. attempted to lead Europe along a path of more liberal tariffs.

(*c*) In both wars government expenditure increased:

(*i*) During the First World War federal government spending in the U.S.A. reached 13 per cent of the gross national product: in Britain government spending grew to more than 50 per cent of the G.N.P.

(*ii*) By the end of the Second World War, in both countries, central government expenditure exceeded 50 per cent of the G.N.P. (By 1950 it was about 20 per cent of the G.N.P.)

Government influence over the economies obviously increased tremendously because of the rise in government expenditure. In addition direct control was exerted to steer production along the lines dictated by wartime needs. The 1939–45 war was followed in Britain by nationalisation of transport, coal-mining, the gas and electricity industries, iron and steel (temporarily) and the Bank of England. Whilst the economic power of government continued to be tremendous in the U.S.A., a direct takeover of industry by the state (except for investment in water resources) did not occur.

CYCLICAL TRENDS

11. 1850–73: upswing.

(*a*) 1850–6. British exports increased, aided by railway development abroad and increasing demand for textiles. American exports of raw cotton expanded; the output of textiles and coal grew; railway construction increased; there was an influx of immigrants. Gold was exported from California and banks operated on a more liberal credit base.

(*b*) 1857. The American financial crisis affected other countries. Britain's economy ran into temporary difficulties. Unemployment was high in 1859.

(*c*) 1861. The American Civil War began and interrupted peacetime economic growth patterns. From 1862 to 1864 Lancashire suffered an acute shortage of raw cotton.

(*d*) 1866. A financial crisis occurred in Britain owing to the failure of Overend & Gurney. Effects were felt in the U.S.A.

(*e*) 1867–73. In Britain recovery from the domestic credit difficulty was rapid after 1867. Both the U.S.A. and Britain enjoyed prosperous business conditions from 1868 to 1873 which were aided by the railway boom in the U.S.A., and the demand for British machinery and railway equipment from the U.S.A. and Europe.

12. 1873–96: falling prices, with fluctuations in economic activity.

(*a*) *The upswing came to an end in* 1873. The downswing was heralded in the U.S.A. by the failure of J. Cooke & Company, who had over-extended themselves in the promotion of railway development. The New York Stock Exchange closed and banks suspended activity. Repercussions were widespread and a depression followed, from which recovery came in 1878–9, aided by exports of grain to Europe, where crops were poor in 1877, 1879 and 1880. Recession set in again in 1884. The downturn in activity in Britain in 1873 was less severe, but difficulties increased in 1877 and 1879–80 because of poor harvests. Recovery commenced in 1889 and lasted until 1893: trade with the U.S.A. was good.

(b) *The setback to growth in economic activity in* 1884 in the two countries was linked with the recession in railway building in the U.S.A. and the fall-off in demand for iron and steel. In Britain shipbuilding activity declined as well: 1886 was a year of great unemployment and the Trafalgar Square disorders. Recession was less marked in the U.S.A. and by 1887 railroad construction, land transactions and heavy and light industries were booming. In the late 1880s, British exports increased, shipbuilding expanded and capital exports to Australia and South America were high.

(c) *Downturn and crisis:*

(i) A downturn in activity began in 1890 in Britain. Exports began to diminish in volume because of the decline in railway construction in South America and Australia especially. Barings had to be assisted by the Bank of England. Difficulties in export-orientated industries and agriculture became acute in 1893: wages were cut. Recovery was slow in developing: activity expanded in 1895.

(ii) In the U.S.A. the business cycle was somewhat different. A crisis occurred in 1893 because of unstable monetary conditions and railroad bankruptcies. The financial crisis deepened into a depression that lasted until 1896, when business recovery began.

13. 1896–1920: rising prices and war.

(a) British exports grew in volume from 1893 to 1896, stagnated for two years and then continued to expand until 1907. Domestic investment increased from 1894 to 1900, and at the turn of the century the economy was on the crest of a wave, with iron, steel and coal prices at high levels. A building boom reached a peak in 1902–3. The fall in iron and steel prices in 1900 did not last long, and setbacks in 1902, 1903 and 1904 were of a relatively minor nature because recessions in different industries were not synchronised. In 1904–5 a rapid increase in exports began; demand for coal was high.

(b) Cyclical trends in the American economy again diverged somewhat from those in Britain. Recovery began in 1896: agriculture enjoyed a favourable shift in the terms of trade; iron and steel output expanded rapidly, but the industry encountered a brief setback in 1900 when iron and steel prices began to fall. Difficulties were brief: recovery commenced in 1901, but in 1903 a financial crisis occurred because of speculative activities on the stock exchanges, culminating in the collapse of several trusts formed for dealing in securities. Business activity diminished but only for a time. Recovery was brisk from 1904.

(c) The difficulties in 1907 originated in the U.S.A., again because of a financial panic, resulting from the collapse of trust companies. Bankruptcies were high, banks failed and a severe decline in business activity set in. British exports to America were affected – trade slumped and shipbuilding and cotton were badly hit, which in turn affected the domestic market. Revival began in 1909 – orders flowed in to British exporters and an export boom was experienced by Britain from 1910 to 1913. Shipbuilding and shipping were prosperous and agriculture reasonably so. Building began to recover. Revival in the U.S.A. was not so confident: it was followed by recession in 1910–11 and then further recovery.

(d) The First World War affected economic trends considerably, but at first in Britain "business as usual" was the motto. As the war lengthened so government intervention increased and blanketed the effect of market forces. Prices, employment and the distribution of raw materials were controlled. The cost of living rose under the impact of inflationary financing of the war. In the U.S.A., after initial pessimistic setbacks, a wartime boom was enjoyed and real incomes increased—but not at a continuous rate. Demands for materials from Europe were high. Government financing of the war was inflationary in nature.

(e) In both countries a post-war boom followed a brief lull in activity after the armistice. Both in the U.S.A. and in Britain the boom was based on rising domestic demand and on exports. The boom was short-lived. Exports fell in both countries, but had a greater impact on the British economy than on that of the U.S.A., where a fall-off in domestic demand was the prime reason for the boom ceasing. Recession set in about June 1920.

14. 1920–39: inter-war boom, depression and recovery.

(a) The American economy experienced a severe recession from mid 1920 to the middle of 1921. Recovery then commenced and from 1922 to 1929 a boom of unprecedented dimensions was experienced (though with setbacks in 1924 and 1927). Prosperity was not spread evenly over industry: it was concentrated largely on the new industries. Coal, textiles, shipbuilding and agriculture had difficulties. The British economy was warped to some extent by the events of 1918–20. An undue proportion of resources remained in, or was newly invested in, the old staple industries, which experienced great difficulty in the 1920s as export markets were lost. Whilst the newer industries became dominant forces in the U.S.A., their growth was too slow in Britain to compensate for the fall in activity in textiles, iron and steel, shipbuilding, coal and agriculture. Whilst the U.S.A. rode high, the British staple industries remained in the doldrums. Unemployment exceeded 7 per cent. Balance of payments problems loomed on the horizon.

(b) The *Wall Street crash* and the subsequent economic depression in the U.S.A. (1929–32) had world-wide repercussions. The U.S.A. virtually ceased exporting capital: Germany, in particular, was badly hit by the withdrawals of capital, and British exporters suffered from the fall-off in world trade. In 1931 Britain abandoned the gold standard: the U.S.A. followed suit in 1933. By 1932 manufacturing output in the U.S.A. was less than a third of the 1929 value, and about 15 million people were unemployed. British exports of coal, textiles, iron and steel and ships fell, and unemployment rose to about 3 million in 1932. Government intervention was necessary in both countries to attempt to effect recovery, which came gradually from 1932 onwards. The recovery in Britain was more successful than in the U.S.A., partly because depression of the newer industries had been less severe: recovery was helped, too, by rearmament after 1935 and by the housing boom. By 1937 manufacturing, construction and public utilities were operating well above the 1929 level of output: industrial production was 20 per cent more than in 1929. In the U.S.A. industrial output in 1937 was about 80 per cent of the 1929 figures.

(c) In 1937 recovery was halted and a sharp recession hit the U.S.A. British recovery was not so badly affected as the American. Industrial output fell by nearly one point: the gross national product fell only slightly, whereas in the U.S.A. it fell by $8 billion. A threatened slide into depression in the U.S.A. in 1938–9 was halted by the onset of the Second World War.

15. The Second World War and the post-war years. Government intervention in economic affairs grew in the Second World War and damped down the play of market forces, but inflation was difficult to curb and the cost of living increased. After the end of the war Britain and Europe found recovery more difficult than

anticipated. Vast amounts of American aid were required
neither the U.S.A. nor Britain, however, did the feared post
depression occur: too much pent-up demand existed for depress
to get a quick toe-hold and international co-operation facilitated
an expansion of world trade.

PROGRESS TEST 1

1. Compare the effects on domestic trade in Britain and the U.S.A. of climate, relief and the size of the two countries. (**1, 2**)

2. Why has government intervention in economic affairs in Britain and the U.S.A. tended to increase? (**5–7**)

3. To what extent was the New Deal successful in facilitating economic recovery in the U.S.A. in the 1930s? (**6**)

4. Consider the short-term and long-term effects of the American Civil War. (**8, 9**)

5. How did the First World War affect Britain's economy? (**10**)

6. Outline the major cyclical trends:

   (*a*) in the U.S.A. economy;
   (*b*) in the British economy. (**11–15**)

7. Why did prices tend to fall in the last quarter of the nineteenth century? (**12**)

8. Compare economic trends in the U.S.A. and Britain in the 1920s. (**14**)

PART TWO

POPULATION

CHAPTER II

POPULATION IN BRITAIN

ECONOMIC SIGNIFICANCE OF AN INCREASE IN POPULATION

1. Economic advantages.

(*a*) A rapidly increasing population, such as the population of the U.S.A. and England before the First World War, has a large proportion of people in the highly productive age-groups: it tends to be flexible and dynamic. A static or declining population tends to have a relatively large proportion of people in the older age-groups and is less dynamic.

(*b*) Investment is stimulated because of the likely expansion in future needs: prices tend to rise and profits are good.

(*c*) The growth of towns is fostered and compact markets increase and make for easier distribution of products and encourage further transport development.

(*d*) A larger domestic market for consumer goods encourages the growth in the size of business units, which tend towards greater economic efficiency.

(*e*) Division of labour is stimulated, giving the advantages of greater specialisation of function.

2. Disadvantages.
A pressure is exerted on capital resources: interest rates tend to be forced upwards. It may be difficult to maintain the capital–population ratio at existing levels. Entrepreneurs and the state may be unable to harness the nation's resources so as to improve living standards.

(*a*) What actually happens depends upon how the nation harnesses its resources to provide for the increasing population. In both Britain and the U.S.A. resources have been harnessed successfully enough to provide better standards of living for most people.

(b) Political and social problems are created too: the nation has to be able to overcome these problems just as much as the economic problems created by an increase in population.

THE GROWTH OF POPULATION, 1750–1951

3. Population before 1850.

(a) Population growth speeded up from about the *middle of the eighteenth century*. From 1066 to 1750 the population of England and Wales is estimated to have increased from about 1 million to 6·5 million, but in the next hundred years the population nearly trebled, reaching a total of 17·9 million. Scotland's population grew from about 1·3 million in 1755 to 2·9 million in 1851.

(b) The *reasons* for the growth in population have been the subject of controversy but the weight of the evidence produced so far appears to favour a lowering of the death-rate as the main factor. However, the death-rate was more than 20 per 1,000 before 1850, and, in fact, increased between 1821 and 1841 because of insanitary conditions in urban areas which resulted in outbreaks of disease such as cholera and typhoid. Obviously a high birth-rate was also a primary factor: from 1811 to 1851 it exceeded 33 per 1,000, which, combined with the death-rate, gave a net reproduction rate in excess of one. Recent research places more emphasis on an increase in the birth-rate.

(c) Population statistics for the eighteenth century and the early nineteenth century must be viewed with *scepticism*. Rickman, Farr, Finlaison and Griffith have produced demographic estimates for the eighteenth century all of which have been criticised for their margins of error. In 1801 came the first census of Britain, but registration of births and deaths was not complete until after 1860, and, again, statistics must be regarded sceptically.

4. Population, 1851–1951.
The following table shows the growth in the population of the British Isles (in millions) between 1851 and 1951:

	1851	1901	1951
England and Wales	17·9	32·5	43·8
Scotland	2·9	4·5	5·0
Ireland	6·5	4·5	4·4*

*N. Ireland and Eire.

The relatively slow growth of Scotland's population and the decline in Ireland's population should be noted.

DEATH-RATES AND BIRTH-RATES

5. Fluctuations in the death-rate, 1850–1950. The death-rate remained in excess of 20 per 1,000 until the mid-1870s and then began to diminish: it was 14 per 1,000 in 1914 and hovered around 12 between 1931 and 1950. Expectation of life at birth increased from forty-three years in 1850 to sixty-six in 1950.

The fall in the death-rate was due to the following factors:

(a) Advances in medical knowledge and the provision of medical facilities.

(b) Sanitary engineering improvements in towns and cities.

(c) Better housing conditions.

(d) Dietary improvements: the availability of a larger variety of foodstuffs improved and so did the knowledge of the effects of diets on the human body.

6. Fluctuations in the birth-rate, 1850–1950. In the 1870s the birth-rate began to diminish significantly. By 1940 it was down to 15·3 per 1,000 and gloomy prognostications were being voiced about a decline in the total population, but in the 1940s the birth-rate increased and pulled the net reproduction rate over one again.

(a) The decline in the birth-rate was due to the following factors:

(i) A more widely diffused knowledge of birth control and a greater availability of contraceptives existed.

(ii) Children became economic liabilities for longer periods as compulsory school attendance was imposed among older age-groups.

(iii) Mass media of communication propagated an increased awareness of the material aspects of life; people kept families small so as to enjoy higher standards of living.

(iv) Many middle-class parents restricted family size so that they would be able to offer better opportunities in education to the children they already had.

(v) Women have become less willing to accept purely domestic roles.

(vi) In the inter-war period a feeling of insecurity resulted from economic changes and helped to keep family size small.

(b) The fall in the birth-rate was most marked among the upper and middle classes between 1870 and 1900, and it spread lower down the social scale in succeeding years, so that in 1939 the number of children living in Britain was 2 million less than in 1914.

(c) The higher birth-rate after 1945 was due to greater economic prosperity, family allowances and better social security. In 1951 the number of children below the age of 4 was 710,000 more than in 1931.

COMPOSITION AND DISTRIBUTION

7. The composition of the population.

(a) Between 1851 and 1911 the proportion of population in the "working" age-groups increased and changes in the birth- and death-rates worked to the economic advantage of the country.

(b) From 1911 to 1951 there was a proportionate increase in the number of people in the age-groups over 65: there was an ageing of the population. The pick-up in the birth-rate after 1945 saved England from a decline in the total population.

Age-groups	Proportion of total population		
	1851	1901	1951
	England and Wales		
20–34	24%	26%	21%
65+	4·6%	4·6%	11%
	Scotland		
20–34	24%	25%	21%
65+	5%	4·8%	10%

8. The distribution of the population.

(a) Between 1871 and 1951 the population of the six major English conurbations increased from 8·35 million to 17 million and that of the Clydeside conurbation from 0·5 million to 1·7 million. Until 1931 the rate of growth of these areas tended to be higher than that of Britain as a whole. After 1931 there was a slight decline in south-east Lancashire.

(b) The proportion of the population living in towns with a total of more than 10,000 inhabitants increased from 39·4 per cent (1851) to 80 per cent (1951).

By 1951 about 50 per cent of the population lived on 5 per cent of the total land area.

(c) The influence of coal and electricity was as follows:

(i) In the nineteenth century the coalfields exerted a pull on industry and population: northern and Midland mining and manufacturing areas increased their population from 7 million to 21 million between 1851 and 1901.

(*ii*) In the inter-war years the generation of electricity increased and made the "newer" industries (radio, television, domestic electrical appliances, motor vehicles, etc.) less dependent on coalfield location. The south-east and manufacturing Midlands exerted a strong pull on population.

(*d*) Population declined in the first half of the twentieth century in fourteen Scottish counties and five Welsh counties: the population of the County of London diminished, too, but that of the London conurbation increased by about 1·8 million.

(*e*) Principal towns experiencing a decline in population between 1911 and 1951 included Blackburn, Bolton, Gateshead, Glasgow, Hull, Liverpool, Manchester, Oldham, Portsmouth, Wigan and Yarmouth. Reasons for this vary:

(*i*) In Lancashire the decline of employment in the cotton textile industry was a basic factor.

(*ii*) The Clydeside conurbation suffered from inter-war setbacks in shipbuilding. In some cases surrounding areas grew at a rate that more than compensated for the decline in city population.

IMMIGRATION AND EMIGRATION

9. Migration of population.

(*a*) *Scope and effects:*

(*i*) Estimates vary. The net outflow of U.K. citizens from U.K. ports in the period 1876–1914 was about 5 million to countries outside Europe.

(*ii*) The aggregate outflow in the period 1853–1914 was approximately 13·7 million, but this ignores inward movements: the net emigration would be less than this.

(*iii*) Probably the net loss of population because of external migration between 1870 and 1914 was about 2 million for Great Britain.

(*b*) *The main reception areas:*

(*i*) The U.S.A. was the main area of attraction to 1905. Emigration, until 1880, tended to flow and ebb with cyclical fluctuations in the American and British economies: in boom times in the U.S.A., emigrants left Britain to settle in the New World; in recessions the flow dwindled and, in fact, at times reversed for brief intervals. After 1880 the coincidence of emigration fluctuations and American economic fluctuations diminished. Emigration to the U.S.A. between 1873 and 1914 was approximately 3 million.

(*ii*) British North America became the most important area after 1905 and absorbed about 1 million U.K. emigrants.

(*iii*) Australia and New Zealand increased in importance as net reception areas after 1905.

(*iv*) South Africa was of relatively minor importance: the main phase of emigration to South Africa was from 1886 to 1905.

(*c*) *Migration, 1914–50:*

(*i*) The First World War ushered in a period of immigration restriction by the U.S.A. so that the average net outflow of British migrants to the U.S.A. fell from 58,000 per year (1907–13) to 30,000 per year (1919–30).

(*ii*) In the 1920s the main net outflow was to Canada, but Australia was also an important reception area.

(*iii*) In the 1930s the net flow was inward. The impact of the economic depression in the western world was the primary cause. Net outflow continued only to South Africa.

(*iv*) After 1945 Australia, New Zealand, Canada and the U.S.A. were the main areas of settlement for emigrants from Britain. From 1946 to 1956 some 800,000 emigrated, of which South Africa received 90,000, but reverse flows developed from other Commonwealth countries.

OCCUPATIONS

10. Occupations, 1851–1951. Industrialisation was accompanied by changes in the employment pattern:

(*a*) Employment in agriculture, forestry and fishing absorbed 21 per cent of the labour force in 1851, but only 4·7 per cent in 1951.

(*b*) The proportion of the labour force in government employment increased from 1·5 to 8 per cent.

(*c*) Commerce, finance and distribution absorbed 5 per cent of the labour force in 1851 and 15 per cent in 1931 but the proportion then declined to 14 per cent in 1951.

(*d*) Throughout the one-hundred-year period manufacturing absorbed between 29 and 34 per cent of the labour force, but within this group textiles and clothing declined in importance whilst metallurgical and engineering industries increased in importance.

(*e*) Since 1931 the number engaged in mining and quarrying has declined sharply.

PROGRESS TEST 2

1. What are the economic effects of an increase in the population of a country? (**1, 2**)

2. Discuss reasons for the increase in population before 1850. (**3**)

3. What effects did changes in the British birth-rate and death-rate have on the composition of the population? (**3, 7**)

4. What determined changes in the birth-rate between 1850 and 1950? (**6**)

5. Examine changes in the geographical distribution of Britain's population in the inter-war period. (**8**)

6. What was the pattern of emigration and immigration between 1850 and 1950? (**9**)

CHAPTER III

POPULATION IN THE U.S.A.

GROWTH AND FLUCTUATIONS

1. The growth of the population.

(a) In 1850 the population of the U.S.A. was approximately the same size as that of Britain; by 1910 it was 92 million and more than double Britain's population; in 1930 it was 123 million; and by 1950 it had reached 151 million – three times the size of the population of Britain.

(b) From the attainment of independence to the Civil War the annual rate of increase was between 3 and 4 per cent, which was almost twice the rate of population increase in England.

(c) After 1860, and until 1900, the rate of increase declined to less than 2 per cent per year: it revived for a decade and then gradually diminished to a nadir of 0·7 per cent in the 1930s, picking up to 1·5 per cent by 1950.

2. Reasons for changes in growth-rates.

(a) *Until 1860 the birth-rate was extremely high:* it is estimated to have been about 55 per 1,000 (much higher than Britain's) whilst the death-rate was 25 (about the same as Britain's). Reasons for the relatively high rates of increase were as follows:

(i) The great abundance of natural resources in the U.S.A. reduced restraint on marriage, which occurred relatively early in life and, therefore, kept the birth-rate high.

(ii) The death-rate was kept relatively low because of better standards of nutrition, and a minimising of contagious disease because of the scattered nature of the population.

(iii) Immigration increased after 1840 and accounted for half the rate of growth by 1850.

(b) *After 1860 there was a fall in the rate of growth* despite the fact that more than 20 million immigrants settled in the U.S.A. between 1860 and 1920. Reasons for the fall were as follows:

22

(*i*) The age at marriage increased: by 1890 less than half the women who married did so before the age of twenty-five, and the average age of marriage for men was twenty-six .

(*ii*) The incentives to have smaller families and to postpone marriage increased as the period of schooling and training for young people lengthened because of the growing complexity of skills needed in industry.

(*iii*) Knowledge of birth-control methods became widely diffused.

(*iv*) Cheap, fertile land became less accessible to the bulk of the population as the frontier extended westwards and, therefore, early marriages became rarer.

(*v*) People deliberately opted for consumer goods and a higher standard of living instead of having large families: the age of high mass consumption dawned, aided by mass marketing and mass advertising.

(*vi*) In the 1930s high unemployment delayed marriages and kept the birth-rate down to 18 per 1,000 because of economic insecurity.

(*c*) *In the 1940s the birth-rate increased once more*, partly because of greater economic security as the economy swung into full production because of the Second World War. Victory in the war was a psychological factor in the increase in the birth-rate after 1945.

COMPOSITION AND DISTRIBUTION

3. The composition of the population.

(*a*) The proportion of people in the age-group over 65 increased and that of people in the age-group under 15 decreased. Proportions were as follows:

	0–14	15–64	65 and over
1850	41%	56%	3%
1900	35%	61%	4%
1940	25%	68%	7%

(*b*) The number of people in the productive age-groups (15–64) increased both relatively and absolutely.

(*c*) Expectation of life at birth rose from thirty-nine years in 1850 to sixty-three years in 1940, but that of the coloured community tended to be less than these figures indicate, although the gap narrowed between 1900 and 1940.

4. The distribution of the population.

(*a*) Urbanisation increased all the time. About 20 per cent of the people lived in urban areas in 1860: in 1950 the proportion exceeded 60 per cent. In 1860 New York had 814,000 inhabitants: in 1900 it had 3·4 million. Chicago's population grew from 109,000 to 1·7 million in the same period. A similar story could be told of other cities.

(b) Whilst the North-east and the South contained the bulk of the population in 1860 and continued to grow, rates of growth were greater in the North Central region and the West after the Civil War.

In 1840 the North-east contained about 40 per cent of the population, in 1940 27 per cent. The population in the South declined, too, from 40 to 32 per cent. On the other hand the North Central region increased its share from 20 to 30 per cent and the West from virtually nothing to 11 per cent.

THE LABOUR FORCE AND SLAVERY

5. The labour force.

(a) The *total labour force* increased rapidly. In 1900 it was 28·5 million; in 1950 it was 64·7 million.

(b) The *pattern of employment* changed:

(i) Agriculture was an important employer of labour throughout the nineteenth century, but its relative significance declined. In 1800 it employed 75 per cent of the labour force, in 1900 about 40 per cent and in 1950 only approximately 11 per cent.

(ii) Agriculture, as a source of employment, was more important both relatively and absolutely in the American economy than in the British economy throughout the period 1850–1950.

(iii) Manufacturing became a more important source of employment. In 1900 19 per cent of the labour force were engaged in manufacturing – approximately 5·5 million. In 1950 23 per cent of the labour force were so employed – about 15 million.

(c) Until the abolition of slavery, *slave labour* was an important part of labour supply, although it declined in proportion to the total labour force.

(d) *Immigrants* formed an important source of labour supply between 1820 and 1920 especially.

6. Slavery.

(a) Slave labour provided about 40 per cent of the labour force in 1800 and only 12 per cent in 1860.

(b) Slavery was confined largely to the South and was associated mainly with agriculture and the plantation system: the number of slaves gainfully employed is estimated to have been 2 million in 1860 out of a total labour force of some 11 million.

(c) Assertions have been made by historians that slavery would have declined because it was economically inefficient: recent research suggests that slavery, in fact, was not unprofitable and that its elimination depended on political measures. Economic forces would not have killed slavery.

(d) Slavery was a primary issue in events leading up to the Civil War between the North and the South.

IMMIGRATION

7. The pattern of immigration.

(a) *Net immigration* was approximately 25 million between 1840 and 1920, and gross immigration in the period was about 35 million. (Net immigration equals gross immigration minus the number of emigrants.)

(b) The *proportion* of the increase in population composed of immigrants from 1840 to 1900 was approximately 25 per cent and from 1900 to 1920 about 30 per cent. After 1920 immigration became a less significant factor in population increase: the *Quota Act* of 1921 imposed a maximum of 358,000 per year, which was reduced further in the 1920s and pegged at 154,000 per year in 1930.

(c) Until 1920 the majority of immigrants were of *European* origin, coming mainly from Britain, Germany and Ireland to 1890, and then mainly from Italy, Austria, Hungary and Russia. After 1910 these last four countries supplied more than 80 per cent of the immigrants.

(d) From 1920 onwards national origins became *more diversified*. Although from 30 to 50 per cent of immigrants were of European origin, Canada, Mexico and Puerto Rico became more important sources.

8. The significance of immigration.

(a) Immigration helped sustain the rate of growth of population, which was an important factor in the high rate of economic growth between 1840 and 1929.

(*b*) More than half the immigrants were in the highly productive age ranges (14–44) and more than half were men. This provided a flow of relatively unskilled labour that was needed to construct railways, to build roads, houses, offices, etc., to man factories and mines, to provide a farm labour force, and to help populate some of the frontier areas. Between 1860 and 1910 more than 50 per cent of the immigrants were unskilled or worked on farms; only 20 per cent were skilled workers.

(*c*) Surges in the growth of national income (*per capita*) to 1929 coincided with periods of high immigration. (However, it does not follow that the rates of economic growth and immigration are linked in the sense that the latter brings about the former: the obverse could and did apply.)

(*d*) Large-scale immigration helped keep down the growing disparity between American and European wages, although wages in the U.S.A. continued to be considerably higher than in Europe, especially among skilled workers.

(*e*) Some writers suggest that immigration provided the U.S.A. with a trained labour force of skilled workers, who were, therefore, a net addition to the country's wealth, but difficulties are encountered in assessing this: many were in need of re-education as the paths of European and American technology diverged.

(*f*) Pressure on resources in Europe was eased by the flow of emigrants to the U.S.A.

(*g*) The effect on wage-rates of immigration varied over time and distance: workers opposed immigration on the grounds that it reduced wage-rates, and in the short term in particular areas it did so, but in the long run immigration did not stop real wages increasing.

(*h*) After 1921 the proportion of skilled and professional workers among immigrants increased in response to the needs of the U.S.A. and because of the imposing of controls over immigration. American technology benefited from the immigration of highly skilled workers, some of whom were displaced persons from oppressive regimes in European countries.

(*i*) British migrants found employment in textiles, mining, metal industries, pottery, glass-making, etc., and British skill was an important factor in the early development of such industries, but, by the end of the nineteenth century, differences between American and British technology were such that the demand for skilled workers from Britain declined severely.

PROGRESS TEST 3

1. Examine the reasons for the changes in the American birth-rate. (**2**)

2. Discuss changes in the composition of the population of the U.S.A. (**3**)

3. How did the distribution of the American population change before 1950? (**4**)

4. Discuss changes in the pattern of employment between 1850 and 1950. (**5**)

5. Consider the changes in the sources of origin of American immigrants after 1850. (**7**)

6. How important was immigration in the growth of the American population? (**7, 8**)

PART THREE

AGRICULTURE

CHAPTER IV

COMPARATIVE ASPECTS

1. Lack of homogeneity. Accurate generalisations are difficult to produce in relation to agriculture because of its heterogeneity caused by:

(a) differences in the nature of soils;

(b) differences in climatic conditions;

(c) varying degrees of economic and agrarian competence among farmers;

(d) differences in the availability of credit facilities, modern equipment and transport facilities among regions;

(e) varying distances from urban markets.

2. Size of the U.S.A. and Britain. The U.S.A. is composed of fifty states, spread over a large area, and has a greater variety of climatic conditions than has Britain: consequently the variety of crops is greater, and generalisations true of one region may not be apt for other regions. The lack of homogeneity is more pronounced in the U.S.A. than in Britain.

3. Surplus production. From about 1850 the U.S.A. has had a surplus of foodstuffs and agricultural raw materials to export. On the other hand Britain has been a food deficit area, relying on large quantities of imports.

4. Relative decline in the importance of agriculture. In both countries agriculture has become relatively less important in its contribution to the national income and in providing employment for labour.

5. Alternation in terms of trade. In both countries there has tended to be a fluctuation in the terms of trade between the agricultural and industrial sectors of the economy, roughly following the pattern of the Kondratiev long-wave cycles in commodity prices, *i.e.* cycles of approximately twenty-five years:

(a) *1850–73*:

(i) Generally, this was a period of prosperity for the agricultural sector of both countries, although rumblings of discontent were heard from British cereal producers in 1850, 1851 and 1852 and American cotton producers suffered severely from the effects of the Civil War.

(ii) American competition in British markets had yet to be felt in all its severity. Britain's imports of foodstuffs were complementary rather than competitive for the home producers.

(b) *1873–96*:

(i) American agricultural production increased by leaps and bounds as more and more land was put to the plough. The rate of growth of output was greater than the rate of growth of demand, resulting in lower prices per unit of output. Between 1870 and 1890 almost 500 million acres (200 m. ha) were added to the area under cultivation. Great unrest existed in American agricultural circles.

(ii) Transport innovations facilitated the expansion of agriculture in the U.S.A. and enabled cheap American produce to be exported to Europe. Britain's free-trade policy left the door open for American agricultural produce to be imported in large quantities. This had adverse effects on British grain producers, who suffered from the resultant fall in income. British livestock producers were affected after 1883, but to a lesser extent.

(c) *1896–1920*:

(i) In the U.S.A. terms of trade between agriculture and industry moved in the favour of agriculture as population increased and the rate of growth of demand for agricultural products caught up with that of supply.

(ii) In Britain economic circumstances improved for the grain producer as the U.S.A. became a less attractive bargain counter for European buyers of foodstuffs. The farmer, too, diversified his output, concentrating more on products in which foreign competition was less effective.

(iii) The First World War meant greater economic prosperity for both Britain's and America's farmers: prices of products and total incomes of farmers increased. The submarine menace made imports difficult and expensive to obtain for Britain, whilst the U.S.A. became a more important supplier of foodstuffs because cheaper sources of supply either were lost because of the war or involved long, risky sea journeys. In both countries the acreage under the plough was increased.

(d) *1920–39*:

(i) The inter-war period, like the last quarter of the nineteenth century, was a dismal one for many farmers in both countries, especially the producers of cereals.

(ii) The state came to the aid of agriculture in both the U.S.A. (with the New Deal policies) and Britain (mainly with a reversal of the free-trade policy and a payment of subsidies) in the 1930s. There was some improvement in economic circumstances between 1932 and 1937.

(e) *1940–50*. The Second World War, like the First World War, changed the fortunes of the industry. Agriculture in both countries emerged from the war in a much healthier economic state than it enjoyed in 1939.

PROGRESS TEST 4

1. Examine the reasons for the heterogeneity of agriculture in Britain and the U.S.A. **(1, 2)**

2. Discuss changes in agrarian prosperity between 1850 and 1900. **(5)**

3. What impact on Britain resulted from the expansion of American agriculture? **(5)**

5. Consider the effects of the First World War on British and American agriculture. **(5)**

CHAPTER V

AGRICULTURE IN BRITAIN

AGRICULTURE BEFORE 1850

1. The years 1815–37. The period 1815–37 was one of some depression in agriculture, although, in fact, depression was confined largely to cereal producers on heavy land, whose costs of production were high. Also the depression was confined in the main to the years 1814–16, 1820–4 and 1833–6. In these years landlords often assisted farmers by allowing rebates off rents, remitting rents altogether, reducing rents or providing help in the form of fertilisers, etc.

2. The Corn Laws. Protection was viewed by grain producers as a means of ensuring agriculture's prosperity by maintaining food prices at a high level. The period was marked by the great controversy about the Corn Laws, especially in the late 1830s and in the 1840s, when the gap between home demand for wheat and flour and home production widened compared with the 1820s and the early 1830s. Porter (*Progress of the Nation*) calculated that from 1841 to 1849 12 to 16½ per cent of the nation relied on imports of wheat at a consumption of 4 to 6 bushels (145–217 litres) of wheat per head per year.

3. Repeal of the Corn Laws.

(*a*) The Corn Laws were repealed in 1846, though a small registration duty remained. The attitude of farmers to the Corn Laws differed according to the nature of their farming. Those whose main outputs were cattle and dairy products, and who had to buy grain as an input, favoured repeal; others did not.

(*b*) The years 1837–42 illustrate the differing attitudes of farmers to the Corn Laws. Harvests were below average but grain prices increased in a greater proportion than yields fell: grain producers were reasonably prosperous. Manufacturers for various reasons suffered from a sharp recession and Cheshire dairy farmers, whose prosperity depended on that of manufacturing areas, suffered a decline in demand for dairy products. The high price of bread was blamed for part of the fall-off in demand and Cheshire dairy producers became active Corn Law repealers.

4. Difficulties, 1847–52.

(*a*) In the years 1847–52 grain producers feared the worst because of the repeal of the Corn Laws: the price of wheat in the period 1849–52 averaged only 40*s*. 11*d*. per quarter compared with 52*s*. 10*d*. for 1842–6 and 66*s*. 4*d*. for 1837–41.

(*b*) Imports increased to over 16 million cwt (802 m. kg) in 1849 and it seemed that imports would determine the movement of grain prices instead of merely following them, but it was not until the 1870s that the full effect was felt.

5. The landowners.

(*a*) *Landownership*. There was a tendency towards a greater concentration of landownership; the large landowners, especially, tended to enlarge their estates. Newcomers to the ranks of substantial landowners were recruited from banking, commerce and industry. Notable among these were Sir Anthony de Rothschild, Alexander Baring and Samuel Jones Lloyd.

(*b*) *Rent adjustments*. Gradually rents had to be adjusted to the lower price levels between 1815 and 1837. Tenants were helped by landowners, who granted rent rebates, rent reductions and remissions, etc. In difficult periods, to some extent, landowners "carried" tenants who were in difficulty. Returning prosperity in the 1840s enabled rents to be adjusted upwards once more.

(*c*) *Government policies*. Landowners benefited in the 1830s and 1840s from changes in government policies:

(*i*) Taxes became lighter.
(*ii*) The *Poor Law Amendment Act*, 1834, reduced the rate burden.
(*iii*) Assistance was provided for landlords to participate in drainage schemes to improve their land.

6. The labourers.

(*a*) *Poverty*. The lot of the farm labourer was not a happy one: the Speenhamland system helped peg wages at a level below subsistence. After 1815 the pressure on wages was downwards because of periods of depression in certain sectors of the industry and because the rate of growth of the supply of labour tended to outstrip that of demand for labour. An accurate assessment of trends is difficult because of variations in allowances in kind and because of the prevalence of the employment of casual labour in many parts of the country. Tooke thought labourers were relatively well off between 1832 and 1836.

(b) *The Poor Law Amendment Act, 1834.* This Act ended the subsidising of some farmers by the ratepayers, but wages still gravitated around the subsistence level, partly owing to the continual growth of rural population.

(c) *Regional variations.* Wages and conditions varied considerably from one region to another: labourers in northern England seemed to have fared better than workers in the south. Caird estimated that average wages in the north were 11s. 6d. a week for a male adult labourer and 8s. 5d. a week in the south, but his statistics are of doubtful reliability. Generally, labourers on large farms fared better than those on small farms.

AGRICULTURE IN THE "GOLDEN AGE," 1852–73

7. Prosperity. The gloomy prognostications of the farmers and landowners, who opposed the repeal of the Corn Laws, were not borne out between 1852 and 1873, when farmers, generally, experienced an age of relative prosperity.

The reasons for farmers' prosperity were as follows:

(a) Lower costs of marketing produce resulted because of the greater spread of the railway network.

(b) An increase in the population of some 5 million people, accompanied by a more than proportionate increase in the national income, due to industrial growth and an expansion of overseas trade, brought about an increase in demand for agricultural produce.

(c) The increased demand for produce pushed up prices and helped increase the total income of farmers. Prices were able to rise, too, because of an increase in the supply of money due to:

   (i) the discovery of new gold reserves in the U.S.A. and Australia; and

   (ii) an expansion of credit facilities and banking in Britain.

(d) Competition from low-cost food producers abroad was hindered by wars (the Crimean War, the American Civil War, the Austro-Prussian War and the Franco-Prussian War) as well as the inadequacy of transoceanic transport.

(*e*) Landowners invested heavily in their estates in spite of the low returns on capital. Social and political reasons accounted for this.

(*f*) Individual farmers were able to increase their returns, partly because of a greater dissemination of knowledge of improved agricultural practices, helped by the burgeoning agricultural societies. (The Royal Agricultural Society of England was founded in 1837.)

8. The organisation of agricultural trade.
Railways revolutionised trade in the agricultural sector:

(*a*) *Local monopolies*. Such monopolies enjoyed by farmers near to large urban conurbations were destroyed.

(*b*) *Urban markets*. There was an orientation of trade towards towns well served by railways.

(*c*) *Regional specialisation*. Trade in dead meat increased: farmers in areas such as Aberdeenshire and Banff were able to become fat-stock producers, selling either directly to buyers in industrial towns and in London or through middlemen in Aberdeen.

9. Landowners' incomes.

(*a*) *Rent increases*. The amount received in rents between 1852 and 1878 increased by about 28 per cent; the greatest increase was in Scotland. Substantial rises occurred in pastoral areas and were reflected in the growing prosperity of dairy farmers and livestock producers. Costs were moving against grain producers before 1873. However, much of the capital invested by landlords was in grain-producing areas; here rent increases were due to interest on the capital expended. The increase in pure rent probably was nil or even negative.

(*b*) *Transport and rents*. Rent increases appear to have been greater near railway stations. In 1863 a Select Committee of the House of Lords estimated that land within five miles (8 km) of a railway station enjoyed rent increases of about 7 per cent because of the proximity of a railway loading and discharging point.

(*c*) *Limited liability*. The development of limited liability enabled landowners to contribute "blind" capital to transport and industrial enterprises from which income was reaped. There was a tendency for landowners to withdraw from direct entrepreneurial activity.

(d) *Mining*. Considerable income was drawn in from mining royalties in coal- and mineral-producing areas, *e.g.* by the Duke of Cleveland, Earl Fitzwilliam, etc.

(e) *Urban rents*. Ground rent and land sales in urban areas produced comfortable incomes for those owning such land. The Duke of Portland had a considerable income from the London area. A Select Committee on Town Holdings (1886) ascertained that very large landowners benefited to a great degree from the growth of urban areas either because of the sale of land or because of ground rent.

10. Labourers in the "golden age."

(a) *Real wages*. Hasbach estimated that the position of agricultural workers deteriorated sharply in the 1850s because of a substantial fall in real wages. The statistics of Bowley, Wood and Rousseau confirm this. There was an improvement in 1870; wages increased by about 7 per cent whilst prices diminished slightly. A sharp improvement occurred between 1870 and 1874 when both wages and prices increased sharply.

(b) *Regional differences*. Accurate generalisations cannot be based on wage and price indices because of regional and local variations in wages and conditions of work, but Steffen calculated that, in terms of bread, the agricultural labourer was $33\frac{1}{3}$ per cent better off in the 1870s than in the 1850s.

(c) *Decline in numbers*. From 1851 the numbers employed in agriculture began to decline. Workers in areas where greater alternative employment opportunities existed would tend to fare better than those in other areas. Labourers in the north of England were better off than those in the south; in Scotland the obverse was true. In Wales the general picture was better than in England. Much employment was of a casual and seasonal nature, and the gang system was rife in eastern and south-eastern counties of England in particular.

(d) *Wage rates as reported to Royal Commission of 1867* (male adult labourers):

    (i) South-east and eastern counties of England: 10s.–13s. a week.
    (ii) Bedfordshire and Buckinghamshire: 16s. 9d.–18s. 9d. a week.
    (iii) Northumberland: 20s. a week.
    (iv) West Riding (plus allowance in kind): 14s. a week.
    (v) Dorset: 8s.–9s. a week.

NOTE: Women earned from 7d. to 8d. a day.

11. Labourers and trade unionism.

(a) *Union growth*. During the 1860s and the 1870s trade-union activity was stimulated by the following factors:

(i) An increase in literacy among farm labourers.
(ii) Rising public sympathy.
(iii) The appearance of Joseph Arch as a leader.

(b) *Unions formed:*

(i) 1865. Midlothian Farm Servants' Protection Society.
(ii) 1866. Border Farm Servants' Protection Association.
(iii) 1866. Agricultural Labourers' Protection Society.
(iv) 1870. West of England Labourers' Improvement Association.
(v) 1872. National Agricultural Labourers' Union, which had 84,000 members by 1874.

(c) *Wages*. The unions were successful in forcing up wage rates by as much as 25 per cent in some cases, but success was short-lived.

YEARS OF DIFFICULTY, 1874–96

12. The Great Depression.

(a) This period has been dubbed as "the Great Depression" but the situation was too complex for a simple generalisation to suffice as an accurate description. Some sections of agriculture were in difficulty. The wheat producer, especially, met with adverse circumstances; to a lesser extent so did other cereal producers and sheep farmers, and to a smaller extent so did live-stock producers after 1883. This may appear surprising in view of the fact that population increased by 10 million, real wages continued to increase and marketing facilities for agricultural produce improved.

(b) Three Royal Commissions took evidence in the years 1881–98:

(i) The Richmond Commission (1881–2).
(ii) The Royal Commission on Depression in Trade and Industry (1884).
(iii) The Second Commission on Agriculture (1894–8).

13. Reasons for the depression.

(a) *Poor harvests*, from 1875 to 1879, caused by inclement weather, were not compensated for by rising prices.

(b) *Prices of cereals* were held down by cheap imports from the U.S.A. and Russia.

(c) *Imports*. A rising tide of imports of grain was facilitated by:

    (i) the free-trade policy of Britain;
    (ii) reductions in the cost of transport because of the more general use of steamships and the continuing spread of the railways.

(d) *Gold supply*. A world-wide change in price levels was caused by changes in the relationship between the supply of and the demand for gold, which became, relatively, a more scarce commodity.

(e) *Wool prices* were affected by the above, and by the growth in production of wool in newly settled areas. (Wool prices rose in 1852–65, declined in 1866–71, fell again in 1872–83: prices in 1883 were less than 50 per cent of prices in 1865.)

(f) *Outbreaks of sheep and cattle disease occurred*.

(g) *The development of refrigerated rail cars* and ship's holds enabled meat imports to increase in the 1880s and 1890s: mutton prices were affected more than beef prices. (Livestock producers, however, benefited from lower grain prices.)

(h) *Cost changes*. Many British farmers were slow to realise the movement of costs against grain production and in favour of meat and dairy production. They did not adapt their methods rapidly enough to combat changing conditions. Farmers were too conservative in their methods of farming.

14. The extent of the depression.

(a) *The Report of the Royal Commission on Agriculture* (1894). This illustrates mounting rent arrears, reductions in rent and an increase in the number of farms in hand.

(b) *Regional variations*. A slide into depression was not true of Britain as a whole: depression was felt most acutely in areas where wheat was the main produce, and where soils were heavy and difficult to work. In such areas labourers and farmers suffered a severe fall in total income without a proportionate fall in costs.

(c) *Prices*. Lower cereal prices were beneficial to farmers who used cereals as inputs – livestock producers and dairy farmers. A beneficial fall in costs occurred. In dairying regions rents generally did not fall: in some cases it was possible to increase them.

(d) *Market gardening*. Horticultural areas near large centres of population enjoyed relatively prosperous times. Competition from abroad was least effective in this sector of the industry because of the perishable nature of the produce.

(e) *Livestock producers* experienced greater competition from abroad after 1883, but the fall in meat prices was by no means as great as that in wheat prices. *Per capita* demand for meat tended to increase, whereas for cereals it diminished slightly.

(f) *Wool producers* were the victims of expanding production in low-cost areas overseas.

(g) *Landowners with alternative forms of income* were able to ride out the depression more easily than those whose sole income was from land.

15. The effects of the depression.

(a) *Decline in arable output.* There was a marked decline in the importance of arable output sold off farms in relation to the total agricultural output.

1873	Arable output = £65m.	Livestock output = £64m.
1896	Arable output = £41m.	Livestock output = £71m.

(b) *Number of livestock* (to nearest million):

	1873	1886	1896
Sheep	29m.	26m.	27m.
Cattle	6m.	7m.	7m.
Pigs	2m.	2m.	3m.
Horses	1m.	1m.	2m.

(c) *The acreage under wheat* declined from 3·6 million acres (1·44 m. ha) in 1874 to 1·7 million acres (0·68 m. ha) in 1896, whilst the oat acreage increased from 2·6 million acres (1·04 m. ha) to 3·1 million acres (1·24 m. ha) and reflected, in particular, the increase in the number of livestock. (Imports of oats grew from 11 million cwt (559 m. kg) per year in the early 1870s to 16 million cwt (813 m. kg) in the later 1890s.)

(d) *The interest of manufacturers* in having low-priced bread available was similar to that of livestock producers: more money was left in the pockets of workers, which:

　　(i) minimised requests for wage increases;
　　(ii) left money over for the purchase of meat and consumer goods.

(e) *Rent arrears.* On estates in cereal-producing regions, where wheat was the main product, substantial arrears of rent built up, and a large number of farms had to be taken over by landowners as notices to quit were sent in.

(f) *Decline of capital investment.* Whilst landowners "carried" agriculture in depressed areas by remitting arrears of rent, reducing rent and providing other assistance for farmers, in many cases less and less capital was invested in buildings, roads, fences and drainage. Capital was repelled by unprofitable farming and attracted away by the lure of more profitable investment opportunities elsewhere.

(g) *Low farming.* Scots found it profitable to move from relatively high-rent areas in Scotland to low-rent areas in East Anglia, Hertfordshire, etc., and to indulge in "low" farming, converting arable into pasture. Alternatively, potato growing was expanded.

(h) *Changes in farming.* There was a considerable increase in market gardening and dairying, partly because traditional farming was no longer so profitable. Increases in real wages and an expanding population in urban areas accompanied by a tendency for an increased *per capita* consumption of fruit, vegetables, etc., encouraged the increase in market gardening.

(i) *Per capita income.* Whereas in 1870 the *per capita* income of those employed in agriculture was about the same as the national average of all incomes, by 1896 the average in agriculture was less than the national average.

(j) *Workers left the industry.* Between 1861 and 1901 the number of male workers diminished by half a million. This created a relative scarcity of labour in agriculture and helped prevent a great reduction in money wages, despite the fall in the total income accruing to farmers and landowners. The labour force obtained a greater share of the total income for agriculture.

(k) *The decline in the relative importance* of agriculture in the national economy continued. By 1901 its share of the national product was less than 6 per cent compared with 18 per cent in 1851; it was no longer the pre-eminent industry.

THE RETURN TO GREATER PROSPERITY, 1896–1920

16. Slow rate of change, 1896–1914.

(a) *Although much had been done to reorganise agriculture* along more profitable lines in accordance with changing conditions, much remained to be done in the period 1900–13.

(b) *Conservatism of farmers.* Agriculturalists tended to be too conservative:

   (i) There was a lack of leadership from landowners.
   (ii) Too much poor farming was tolerated.
   (iii) Management tended to be less efficient than was required if the industry was to be internationally competitive.

(*iv*) Standards of education and training among workers and farmers were low.

(*c*) *Value of output.* The decline in the value of output was arrested. Between 1896 and 1913 annual output increased in value by about £4 million. The decline in rents was halted too.

(*d*) *Wages of labourers* rose slightly and roughly kept in step with the mild increase in the cost of living. Real wages, therefore, did not increase. Labour's share of the total agricultural income declined slightly, despite an increase in the labour force between 1901 and 1911 of approximately 100,000.

17. Government intervention, 1914–20.

(*a*) *Processes of change.* The cost-push away from arable farming and cereal production, to the production of livestock, dairy produce etc., was reversed by deliberate government policy towards the end of the First World War.

(*b*) *Increase in cereal production.* U-boat activity caused great losses of British merchant ships and the government planned an increase in cereal production from British farms to help minimise bulk imports. Between 1916 and the end of 1918 3 million acres (1·2 m. ha) were added to the area under the plough; wheat production was increased by about 60 per cent, that of oats by nearly 50 per cent and potatoes by approximately 40 per cent, whilst the output of milk and wheat fell. In terms of calories, food production rose by 24 per cent.

18. The Corn Production Act, 1917.

(*a*) *Minimum prices.* For the next three years minimum prices for grain were guaranteed to encourage grain production. (However, high prices rendered this provision inoperative.)

(*b*) *Rents.* Landlords were prohibited from raising rents.

(*c*) *Wages.* Agricultural labourers were guaranteed minimum wages. An Agricultural Wages Board was set up to fix wages.

19. The Agriculture Act, 1920.

(*a*) Guaranteed minimum prices for grain were to continue, but there was to be an annual review of the guaranteed prices.

(*b*) The Agricultural Wages Board was to continue fixing legal minimum wage rates.

(*c*) Compensation for disturbance had to be paid to farmers who were evicted by landowners, unless the eviction was for negligent farming. (Compensation was still payable for improvements under the Acts of 1875, 1883, 1900 and 1906.)

THE INTER-WAR YEARS AND DEPRESSION: THE REVERSAL OF LAISSEZ-FAIRE POLICY

20. The course of prices.

(*a*) 1920–3: Rapid fall.

(*b*) 1926–30: Steady decline.

(*c*) 1930–3: Rapid fall.

(*d*) 1933–9: Recovery at a slow rate.

21. Effects of the fall in prices.

(*a*) *Minimum price guarantees.* The drastic fall in prices during 1920–1 put the government in the position of having to provide large subsidies for farmers under the minimum price guarantee. In 1921 the *Corn Production Repeal Act* was passed: it ended, temporarily, attempts to guarantee price levels for agricultural produce and the control of minimum wages.

(*b*) *A decline in arable farming* and grain production occurred.

(*c*) *Comparative advantages.* Greater concentration occurred on products for which comparative advantages were more pronounced – milk, eggs, vegetables, etc.

(*d*) *Fall in value of land.* Landowners became more willing to sell farms (to some extent this was influenced by taxation, too).

(*e*) There was a *reduction in the number of people employed* on the land and an increase in mechanisation occurred, especially after 1933.

(*f*) *Productivity* of those still employed increased.

22. Minor changes in policy.

(*a*) *The Agricultural Wages (Regulation) Act*, 1924, restored state regulation of agricultural wages.

(*b*) A *subsidy* was granted for the producers of sugar-beet under the *British Sugar Subsidy Act*, 1925.

(*c*) Agricultural land and buildings were freed from local *rates* burdens in 1928.

(*d*) The *Agricultural Mortgage Corporation* was established in 1928 under the *Agricultural Credit Act*.

23. Major changes in policy: protection. The depression of 1929–32, a growth of economic nationalism abroad, pressure from the Dominions and strategic motives prompted major upheavals in Britain's economic policies. Agriculture was protected and aided by tariffs, import quotas, subsidies, efforts to regulate production and efforts to fix prices under various Acts.

(a) The *Agricultural Marketing Acts* of 1931 and 1933 enabled boards to be set up to regulate the marketing of milk, hops, pigs and potatoes.

(b) The *Wheat Act* of 1932 provided for the payment of subsidies to facilitate the increase in wheat production to a level of 27 million cwt (1,372 m. kg), later increased to 36 million cwt (1,829 m. kg).

(c) *Import duties* were imposed on horticultural produce and on oats and barley under the *Import Duties Act*, 1932.

(d) The *Cattle Industry Act*, 1934, and the *Livestock Industry Act*, 1937, enabled subsidies to be paid to livestock producers.

24. The effects of protection.

(a) *Wheat.* The acreage increased.

(b) *Foodstuff prices* rose in the shops: estimates of the annual costs of state aid and higher prices resulting from protection vary from £32 million to £100 million.

(c) *Price index.* Despite increases in prices the agricultural price index in the 1930s did not reach the 1929 level. Farmers had great difficulty in operating profitably.

(d) *Drift from the land.* Protection did not halt the drift from the land. By 1939 less than 4 per cent of the labour force was employed in agriculture and forestry.

(e) *Although total production increased*, the extent of the increase is difficult to judge with certainty. Controversy exists as to whether production reached the level of 1913.

(f) *The decline in the total arable acreage* was not halted. Livestock and livestock products accounted for more than 70 per cent of the value of the total agricultural output in 1938.

(g) *Milk production* and consumption was stimulated by the Milk Marketing Board, but the most successful of the marketing schemes was that for potatoes.

(h) *Low incomes.* Whilst protection injected some prosperity into farming, it failed to lift it out of the slough of depression. Rents remained low, farms were difficult to let and many farms were under-capitalised. Many areas in 1939 bore the stamp of neglect.

FARMING DURING AND AFTER THE SECOND WORLD WAR

25. Effects of the Second World War.

(a) *Agriculture*, as in the First World War, became a strategic industry, essential to the nation's survival.

(b) *Arable increase.* The main feature was the increase in arable area and the output of cereals. Between 1939 and 1955 the output of grain sold off the farm trebled, and the acreage increased by about 50 per cent.

(c) *Increased yields* were obtained by a greater use of fertilisers.

(d) *Increased production* was obtained by the use of County War Agricultural Committees, which determined land use, allocated labour and materials and gave advice, and by the use of certain financial inducements:

    (i) Subsidies for ditching, drainage schemes, etc.
    (ii) Subsidies for ploughing up permanent pasture.
    (iii) Acreage payment for wheat and potatoes.
    (iv) Loans for agricultural requisites.
    (v) Subsidising of fertilisers.
    (vi) Guaranteeing prices.

(e) *Agricultural incomes* rose rapidly. The industry was in a healthy state of prosperity in 1945.

(f) The *Agriculture Act*, 1947, was passed to ensure stability for Britain's farmers:

    (i) Prices were guaranteed at a level fixed by annual review.
    (ii) The National Agricultural Advisory Service set up in 1946 was continued, as were County Committees which still possessed the power to control agriculture.
    (iii) An increase of 20 per cent in output was called for to save imports (see C. H. Blackburn, "Import Replacement in British Agriculture," *Economic Journal*, 1950).

PROGRESS TEST 5

1. Outline reasons for the controversy about the Corn Laws and their repeal in 1846. (**2, 3**)

2. Why was British agriculture prosperous between 1853 and 1873? (**7**)

3. What were the sources of landowners' income in the "golden age"? (**9**)

4. How prosperous were agricultural labourers in the "golden age" of farming in the nineteenth century? (**10**)

5. Why was there an agricultural depression after 1875? (**12, 13**)

6. Point out regional differences in the depression in the last quarter of the nineteenth century. (**14**)

7. Examine reasons for and the effects of changes in British agricultural tariffs after 1930. (**22, 23, 24**)

8. What impact did the Second World War have on British agriculture? (**25**)

CHAPTER VI

AGRICULTURE IN THE U.S.A.

THE EXTENSION OF THE FRONTIER

1. The role of the federal government. Between 1781 and 1867 the federal government obtained, by various means, some 1,840 million acres (736 m. ha) of land, much of which was disposed of to state governments, railway and canal companies, private citizens, etc. The objects of public land disposal were:

- (*a*) to raise revenue from land sales;
- (*b*) to promote the settlement of the West;
- (*c*) to facilitate transport development;
- (*d*) to encourage states to develop their resources;
- (*e*) to promote the improvement of agricultural education.

2. Area under cultivation. There was a phenomenal increase in the area under cultivation from 1860 to 1910.

- (*a*) 1860–70. 500,000 acres (200,000 ha) were added to the area under cultivation.
- (*b*) 1870–80. 190 million acres (76 m.ha) were added (approximately the area of France and Britain combined).
- (*c*) 1880–90. 330 million acres (132 m. ha) were added.
- (*d*) 1900–10. 40 milloin acres (16 m. ha) were added.

3. Effects of territorial expansion. The expansion of territory had profound economic, social and political effects:

(*a*) *Speculation.* Although public land-disposal policy became more liberal as the nineteenth century passed, it favoured the buying of land by speculators rather than by settlers. Absentee ownership grew, especially with the granting of land to railroads (some 200 million acres (80 m. ha) were granted to railroad companies).

(*b*) *Output.* There was a vast expansion of agricultural output and a rapid increase in agricultural productivity.

46

(c) *Per capita output.* Emphasis upon output per head in agriculture was greater than the emphasis upon output per acre because of the supply relationship between the two factors land and labour.

(d) *Foreign capital* was attracted to finance the development of resources.

(e) *Population* migrated to the West.

(f) *A large consumer market* developed in the West and new industries arose, resulting in a tremendous expansion of domestic trade.

(g) *Surplus production.* The rapid rate of growth of world agricultural output from 1864 to 1897 acted adversely on farmers' prosperity. This fostered the growth of farmers' movements and resulted in pressures for inflationary policies to be followed to lighten the load on the debtor western areas. (Economic conflict between East and West ensued – although some eastern groups favoured inflationary policies, too.)

(h) *Exports.* Agricultural commodities were important factors in the export trade of the U.S.A. as the country became a food surplus area. Between 1814 and 1914 the U.S.A. progressed from debtor on current and capital account to creditor on current account and then creditor on capital account. Agriculture played a significant role in this growth of the U.S.A.

4. The frontier as a safety valve. Professor Turner has postulated that the existence of frontier areas in the U.S.A. fulfilled the function of a safety valve, allowing surplus labour an avenue of escape to the West away from the subordination to others in the industrial East.

(a) *Contrary views.* Modern economic and social historians dispute this hypothesis. In fact recent researchers have postulated that the industrial towns were more in the nature of safety valves for labour than were the frontier areas. Industrial workers tended to enjoy more regular employment and to receive higher wages than agricultural workers.

(b) *Capital requirements.* Before the *Homestead Act* of 1862 the accumulation of the capital needed to buy and settle on a reasonable farm was beyond the reach of the vast majority of unskilled workers, and the 1862 Act, whilst enabling a homestead to be settled free of land charges, did nothing to provide for the other considerable outlays required in setting up a farm.

Effective choice by labour of self-employment in agriculture or wage earning in industry was not all that great.

(c) *Ebb and flow of migrants.* Statistics of migration to the West tend to show that movement from the East to the West was greater in times of relatively high industrial employment and less in times of relatively high unemployment. This is the obverse of what would be expected if the West had been an efficient "safety valve."

5. The psychological impact of the frontier.

(a) *Profit motive.* Frontier conditions created an environment favourable for profit making and for profit seeking. Speculative land purchase resulted: farmers bought land more from the speculative motive than for actual farming.

(b) *Booster mentality.* It is feasible to postulate that a "booster mentality" resulted from successive stormings of frontier barriers. The farming frontier, the mining frontier, the transport frontier, the technological frontier, etc., were pushed further away as one advance succeeded another. At one and the same time agricultural output grew tremendously and became more diversified, natural mineral resources were exploited, transport innovations developed, industry coalesced into bigger units of a complex character so that they were ready for mass production to cater for uniform tastes that were, to some extent, standardised by the impact of the frontier, which gave rise to massive demands for firearms, barbed wire, farm machinery, sewing machines, etc. It is feasible that frontier areas became good markets for standardised products and played a significant role in the growth of mail-order stores. Certainly the conquering of successive frontier barriers seems to have imbued American society with great self-confidence and, therefore, to have provided a good psychological base for rapid economic expansion.

AGRICULTURE BEFORE THE CIVIL WAR

6. Regional specialisation. Whilst the great diversity of crops were produced, some regional specialisation developed, determined to a great extent by climatic conditions.

(a) *Tobacco* became a main crop in Maryland, Virginia and the Carolinas.

(b) *Cotton* was king in Mississippi and Alabama and produced in great quantities in Texas, the Carolinas and Georgia by 1860. The export market was of great importance, and there were strong links with Britain's Lancashire cotton industry.

(c) *Market orientation in the North.* In the northern states agriculture was orientated more to the home market and the rapidly growing urban conurbations. After the opening of the Erie Canal and increasing competition from farmers further west, there was increasing concentration on dairy farming, fruit growing and truck farming in the East.

(d) *West of the Alleghenies* maize and wheat were the main crops, wheat being the export crop and maize the animal feeding crop. Pigs fed on maize were sold in Cincinatti and Louisville, important meat-packing centres.

(e) *Around the Great Lakes* farming was diversified: wheat was important and so too were cattle and hogs.

(f) *The Mid-West* was a food surplus area and the South a food deficit area: the Mississippi was an important artery down which foodstuffs flowed to southern markets.

7. The cotton-producing South. The perfection of the cotton gin and the expansion of the British cotton industry enabled cotton to become the most important crop in the South. Profligate use was made of land: a form of soil-mining occurred. As one area of land diminished in fertility, so the cotton planter abandoned it and moved to more fertile land elsewhere, so that by 1860 cotton growing was spread over a wide area. Production, at times, grew at a greater rate than demand and led to a search for ways and means to break through cost barriers. Four methods were used:

(a) Migration to more fertile areas.

(b) Diversification of output to reduce outlays on foodstuffs, etc. (diversification was greater than some writers indicate).

(c) The adoption of improved practices.

(d) An increase in the size of the farming unit (some plantations were extremely large.)

8. Slavery. Slaves were used on cotton, tobacco and sugar plantations. Controversy exists as to whether slavery collapsed because it was an inefficient, unprofitable form of organisation. The conclusion of recent researchers is that the system was profitable and that it ended because of other considerations, but other historians, assert that the system was uneconomic and was doomed to failure because it was unprofitable.

9. Grain production.

(a) *Wheat*. This grain was grown over large areas, including the South. Between 1840 and 1860 total production more than doubled: the main producing areas at the latter date were Illinois, Indiana, Wisconsin and Ohio. High-yielding wheat areas experienced highly commercialised farming – large units, absentee ownership, tenant farmers, hired labourers, considerable investment in implements and horses. The area of production moved westwards and an immense flow of wheat from west to east resulted. Wheat, like cotton, was an important export and an important item in the balance of trade. By 1860 more than 6 million cwt (30·5 m. kg) per year were being exported to Britain.

(b) *Corn*. This was very widely grown as animal foodstuff: it was important in the South as well as in the North and West.

10. Livestock.

(a) *Livestock were an essential part of farming* over a very wide area, but main areas of production for the various animals were as follows:

> (i) *Hogs*. Ohio, Iowa, Missouri.
> (ii) *Beef cattle*. Texas, Illinois, California.
> (iii) *Sheep*. Ohio, Texas, California, New Mexico.

(b) *Feeder regions developed* in the West for fattening cattle ready for driving to the East. The coming of railroads transformed the cattle market; specialisation was enhanced, and costs of transportation fell. By 1855 the Erie Railway was transporting to the New York market each year 100,000 head of cattle, more than 200,000 sheep and 150,000 hogs.

(c) *The development of Chicago* as a rail terminal pulled the meat trade to the city, which became the premier meat-packing centre, ousting Cincinatti and St Louis.

11. Agricultural implements.
A new industry for producing agricultural implements and machines was established by 1850. Special attention was focused on ploughs to break up the virgin earth, so that by 1850 American ploughs were ranked best in the world. Important developments were:

> (a) the John Deere steel plough which could break up the tough Prairie turf;
> (b) the Cyrus McCormick reaper;

(c) the introduction of transportable steam-threshing machines;

(d) the development of efficient grain drills.

By 1860 the production of agricultural machinery was a major
industry with total production valued at $34 million. Technology
had prepared agriculture for a great leap forward.

12. The agricultural labour force.

(a) *Agriculture provided employment* for the majority of the
labour force of the U.S.A. in the first half of the nineteenth
century. About 75 per cent of the labour force was so employed
in 1800 and 50 per cent in 1860. In the South, the trends were not
quite the same: the proportion of the labour force in agriculture
exceeded 80 per cent throughout the half-century.

(b) *Wages fluctuated* from area to area, as did regularity of
employment. Statistics are too unreliable to reveal the actual
changes in real wages, but they appear to have increased between
1800 and 1850.

(c) *Much payment was made in kind.* Many labourers lived in
with the farmer and payments were made at irregular intervals.

THE AGRICULTURAL EXPANSION, 1860–1914

13. Four changes of great importance. These were as follows:

(a) *The system of slavery ended.*

(b) *The "last frontier" was reached*, territorially.

(c) *Agriculture was subject to intensive mechanisation*, increasing total production and productivity by leaps and bounds.

(d) *Industry overtook agriculture* as the major contributor to
the national income.

14. The impact on the South of the Civil War and the end of slavery.

(a) *Prostration of agriculture.* In the South plantation agriculture
was almost prostrated during the Civil War: there was a virtual
cessation of the cotton trade. In 1861 4 million bales of cotton were
produced: from 1862 to 1865 only 3 million bales were produced.
Some land was taken out of cultivation.,

(b) *Slave owners suffered capital losses* when slaves were declared
free. To overcome labour problems sharecropping was resorted to
more frequently.

(c) *Credit facilities.* There was a lack of credit facilities in the South: the gap was filled by merchants and storekeepers, who developed the crop lien system under which money was advanced to farmers, crops being the security. Cotton was a favourite crop for suppliers of credit: this helped cause over-production of cotton after 1877 and a considerable fall in land values occurred.

(d) *The crop lien system* accounts to a great degree for the neglect of banking and for the failure of the South to build up a sound structure of credit and finance. Sharecroppers lacked the capital for investment in their land: they needed the support of a sound financial system. Many were swindled out of their dues by moneylenders and book-keepers.

15. The reaching of the last frontier.

(a) *The number of farms* increased from 2 million to 6 million in the period 1860–1910. At least 50 million people were supported by agriculture in 1910.

(b) The expansion is accounted for by various factors:

(i) *Gold rushes.* Some of the would-be gold-diggers involved in the gold rushes in California, Oregon, Colorado, etc., found it more profitable to become farmers.

(ii) *Droughts* in the Ohio valley caused a westward migration.

(iii) *Railways* opened up land for cultivation and the companies actively encouraged immigrants from Europe to settle in the newly developing areas.

(iv) *Technological improvements* made possible the exploitation of the Prairies. Ploughs, harvesters, grain drills, harrows, threshing machines and reapers were needed on a large scale: industry was able to provide them. The general use of more sophisticated machines increased productivity in agriculture between 1860 and 1900 approximately twentyfold.

(v) *Transport innovations* enabled American farmers to compete on favourable terms in European markets with European farmers. Costs of transport fell dramatically: surplus produce was exported in vast quantities.

(vi) *Public land disposal policy* became more liberal. The *Homestead Act*, 1862, enabled individuals to claim by settlement 160 acres (64 ha) of land. (In fact speculators reaped considerable gains from this Act: homesteaders obtained only $3\frac{1}{2}$ per cent of the land west of the Mississippi.)

(vii) *A Department of Agriculture* was set up by Congress in 1862 to (1) promote agricultural research and (2) distribute information about agriculture to agriculturalists and thereby improve the quality of products.

(*viii*) The *Morrill Act*, 1862, facilitated the establishment of "land-grant agricultural colleges" and in 1867 the *Hatch Act* provided for federal assistance for agricultural experimental stations.

(*ix*) *An influx of British capital* helped build up the western range-cattle industry. Companies concerned with the cattle industry numbered among them the Anglo-American Cattle Co. Ltd., the Prairie Cattle Co. and the Colorado Ranch Co. In 1900 only eight out of thirty-seven such companies survived.

16. The mechanisation of agriculture.

(*a*) Between 1860 and 1914 great improvements were made in agricultural machinery and implements:

(*i*) *A self-binder* for harvesting grain had been perfected by 1880.

(*ii*) *Threshing machines* were subject to considerable improvement, increasing productivity more than threefold.

(*iii*) *Steel ploughs* capable of ploughing two, three and four furrows appeared.

(*iv*) *Seed drills* were much improved.

(*v*) *Steam engines* were used for grinding grain, cutting turnips, etc.

(*vi*) *Steam tractors* appeared, but were not in general use.

(*b*) *Mechanisation reduced production costs* considerably in the case of wheat and oats, increased the area one man could deal with successfully, and stimulated the expansion of the agricultural machine industry. In 1910 machines were cheaper and more efficient than those of 1860.

(*c*) *Lower production costs* and increased productivity plus the tremendous territorial expansion in cultivation made the U.S.A. a bargain counter for foodstuffs between 1860 and 1900. This had the effect of stimulating foreign demand for American produce, and facilitated an increase in home demand for industrial goods because of a lowering of the cost of living.

THE DECLINE IN THE RELATIVE IMPORTANCE OF AGRICULTURE'S CONTRIBUTION TO THE NATIONAL PRODUCT

17. Increase in national output and agricultural output.
Whilst agricultural output increased more than threefold between 1860 and 1914 (in terms of constant prices) the industry's proportionate contribution to the national product diminished from 32 to about 20 per cent. Whilst agriculture was expanding its production

other sectors were expanding more rapidly. The total national product increased by more than 500 per cent in the same period.

(a) *The number employed* in agriculture increased from 6·3 million in 1860 to 11·49 million in 1907: after 1912 the number was less than 11 million. Between 1860 and 1914 the proportion of the nation's labour force employed in agriculture declined from about 60 to 25 per cent.

(b) *Only in the North Central and the western states* did agriculture keep pace, relatively, with the growth of the national product, and even here there was a tendency to fail to keep pace after the turn of the century.

(c) *In the spheres of exports* agriculture was of dominant importance to 1900: between 1860 and 1900 agricultural products accounted for approximately 76 per cent of all exports: 1900, relatively, was the peak year, when 80 per cent of exports consisted of agricultural products. By 1914 industrial products formed a greater proportion of exports than agricultural products (in value): *see* **18** below.

18. Agricultural and industrial products as exports. Agricultural products failed to keep pace with industrial products as exports for various reasons:

(a) *Domestic consumption* of foodstuffs tended to increase at a faster rate than supply after 1896.

(b) *Tariffs.* Europe (especially Germany and France) used tariffs to protect its agriculture from American competition.

(c) *The growth of large-scale units* in industry facilitated the minimising of costs, enabled factory production to expand enormously and cater for mass markets on a huge scale at relatively low prices. This boosted the exporting of manufactures.

UNREST IN AGRICULTURE

19. Causes of unrest.

(a) *Total farm income* fluctuated and tended to fall in the 1870s; it rose in the 1880s and changed little in the 1890s and rose substantially in the period 1896–1914. The fortunes of individual farmers varied according to the nature of the products sold off the farms and individual market responses. Generally, however, supply-demand relationships in wheat or livestock were affected by the expansion of agriculture, internationally, between 1860 and 1896, and the American domestic market experienced sharp variations in price: there was a tendency for supply to outreach demand, especially in the 1870s. At the same time price levels tended to fall because of the relatively slow rate of growth of the world's supply of money.

(b) *Prices of agricultural commodities*, on the whole, fell less than those of manufactured goods, which also improved in quality. The terms of trade for agriculture did not deteriorate over the whole period 1865–96, but they did decline somewhat in 1867–70, 1876–9 and 1882–6.

(c) *Discrimination*. Especially in the West North Central states farmers complained of unfair discrimination against them by railroad companies, although there was a reduction in railroad rates after 1874. There was substance in the farmers' complaints against railroad companies in relation to freight charges, freight facilities, the grading of wheat and the grading of corn, but there was some improvement in narrowing the gap between the average selling price of wheat in the U.S.A. and that received by farmers of the West North Central states between 1871 and 1915.

(d) *The crop lien system* in the South put many small freeholders in danger of forfeiting their farms to storekeepers because of their inability to pay for goods advanced on credit against crops. In the North Central areas complaints were made about the imposition of high interest rates for mortgages because of imperfection in the capital market.

(e) *Price fluctuations*. Perhaps the main cause of unrest was the fluctuation in market prices that occurred, and the failure of prices to respond to local supply conditions. Poor harvests might be accompanied by low prices, the obverse of what could have been expected before 1860. International market conditions affected some agricultural prices: these could move along different curves from those that would be expected in a particular locality. Price movements were difficult to predict.

THE FARMERS' MOVEMENTS

20. The Granger Movement.

(a) *Aims*. This movement was social and educational as well as economic in its aims. It originated in 1869 in Washington D.C. and by 1875 had 850,000 members – mainly from farmers' parties

which flourished at the same time and which were political in
nature.

(b) *Granger Laws*. Between 1867 and 1874 a number of state
laws in Illinois, Wisconsin, Minnesota and Iowa effectively
checked the monopoly powers of the railroad companies. The
states claimed regulatory powers over such companies: these
principles were upheld by the Supreme Court in a series of decisions in 1877. The laws were known as "Granger Laws" because of
the part played by the farmers' parties in securing enactments.

(c) *Farmers' co-operatives*. The Granger Movement was involved
in the setting up of farmers' co-operatives for buying supplies,
selling farm products, venturing into meat packing, flour milling,
etc., but the co-operative ventures were short-lived.

(d) *Decline*. The successes in achieving a curbing of the railroad
companies' powers caused a fall-off in interest in the farmers'
parties, and the Granger Movement declined rapidly after 1874–5.

21. The Greenback Movement.

(a) *Chief supporters* of the movement were people from debtor
areas—the South and the North Central states. In effect they
requested a managed currency that would change in volume
according to the needs of business.

(b) *Reasons* for the rise of the Greenback Movement were as
follows:

(i) The rate of growth of the world's gold supply did not keep pace
with demand. This had a deflationary effect that had dire results for
those with large debt burdens.

(ii) The $450 million in American notes (greenbacks) authorised
during the Civil War was the main legal tender until 1879 when the
dollar returned to the gold standard, but the volume of greenbacks
fluctuated well below the 1862 level until the amount was fixed in
1878. The followers of the Greenback Movement desired an increase
in the volume of American notes to lighten debt burdens and to
compensate for contractions in the volume of National Bank notes.

(iii) National Banks were permitted to issue notes under the
National Bank Act, 1863: an Act of 1865 gradually drove other note
issues out of circulation by the imposition of a 10 per cent tax. However, National Bank notes tended to fluctuate in volume contrary to
the needs of business. A managed currency was regarded as a means of
overcoming this deficiency.

(iv) The *Resumption Act*, 1875, aimed to reduce the volume of
greenbacks from $382 million to $300 million. This measure roused
much support for the Greenback Movement among farmers in the

North Central states and the South. The protest movement press-
urised Congress into amending the Act in 1878, fixing the volume of
greenbacks at $346,681,016.

(c) *Decline.* The Greenback Movement fizzled out in the 1880s
after a brief moment of promised growth in the election of 1878.

22. The Free Silver Movement.

(a) *Money supply.* The Free Silver Movement stemmed from
the desire to increase the volume of money in circulation. Supporters were the silver-mine owners, who deplored the Act of 1873,
which effectively de-monetised silver, and farmers (especially
the Southern Alliance and the North-western Alliance Movements).
Opponents were gold-standard advocates who suggested that free,
unlimited coinage of silver would cause unlimited inflation.

(b) *Results of the Free Silver Movement:*

(i) The *Bland-Allison Act* was passed in 1878 authorising the
Treasury to buy silver at the rate of $2 million to $4 million a month
to coin into dollars, exchangeable for silver certificates: the volume of
silver purchases was to fluctuate according to the state of trade. (In
fact the minimum was adhered to.)

(ii) The *Sherman Silver Purchase Act* was passed in 1890. Monthly
purchases of silver were fixed at $4·5 million to be bought with Treas-
ury notes, which were redeemable in either gold or silver but which,
in fact, were generally redeemed in gold. Since silver was to be
minted only to replace or retire such notes very little silver coinage
occurred. Silver bullion simply accumulated in Treasury vaults: it
did not add to money supply beyond the effect of Treasury notes for
purchase of silver in the first place.

23. The Populist Movement. The Populist Party was formed at a
convention in St Louis in 1892. Its demands were as follows:

- (a) Free coinage of silver.
- (b) Greenbacks to replace National Bank notes.
- (c) A currency supply in circulation of at least $50 per head of
  the population.
- (d) A postal savings system.
- (e) Economy in government spending.
- (f) A universal civil service system.
- (g) Improved conditions for labour.
- (h) Control of immigration.

The Populists appealed to a broad spectrum of the community
and increased in strength between 1892 and 1894, but then with-
ered away, especially after fusing with the Democrats in 1896.

Probably increased agrarian prosperity after 1896 killed farmers' interest in the Populist Party.

24. The Farmers' Union.

(a) *State charter.* This movement received a state charter of Texas in 1902. It aimed to increase farm incomes by controlling the marketing of farm produce, but it did indulge in political lobbying in Washington and state capitals for:

  (i) direct government loans to farmers;
  (ii) limitations on land monopolists;
  (iii) state control of railroads and of mining resources.

(b) *Southern origins.* Although the Farmers' Union originated in the South it spread its activities into the states of Illinois, Kansas and Missouri.

25. The American Society of Equity. This society was very similar in its aims to the Farmers' Union and, in fact, merged with it in 1934. Its primary aim was to regulate the marketing of products in the Mid-West and thereby minimise price fluctuations. Each state group of the society tended to pursue independent policies; the Kentucky group achieved notoriety for its militant opposition to the American Tobacco Company.

GEOGRAPHICAL CHANGES

26. Geographical shifts, 1860–1914.

(a) *Grain production* tended to move westwards, even in the early years of the twentieth century. Hard wheats were favoured for bread making and these grew best in the regions of limited rainfall – in the Dakotas, Kansas, Nebraska, etc. Eastern states, unable to compete with the lower-cost areas of grain production in the West, turned to dairying, fruit growing, truck farming, livestock production, etc., where such activities were feasible because of local cost advantages or marketing advantages.

(b) *The corn belt* also moved westwards (and northwards) as did the main areas of oat growing.

(c) *The beef-cattle industry* declined in the North-east and expanded on the Great Plains, the Prairies, the Pacific coastal regions and on the central plains of the cotton belt. By 1900 Chicago was the centre of the most important dairying region of the U.S.A.

(d) *Hog production* was widely scattered, but some concentration in the areas of large-scale corn production was noticeable.

(e) *Sheep rearing* diminished on the eastern seaboard and increased in the West – especially west of the Mississippi.

(f) *Virginia, Maryland and Ohio declined in importance* as tobacco-growing regions. North and South Carolina increased in importance.

(g) *Texas* became the chief cotton state.

AGRICULTURE, 1914–20: THE EFFECTS OF THE FIRST WORLD WAR

27. The increase in European demand for American produce. European demand for American produce increased because:

(a) the Allied powers' imports from Baltic countries and Russia were cut off by the German navy;

(b) long sea routes to Australia and South America were more hazardous than the shorter sea routes to the U.S.A.;

(c) Europe's output of foodstuffs diminished; supplies were sought in the most convenient supply market.

28. The effects of the increased European demand for foodstuffs.

(a) Exports of foodstuffs increased from an average of 6·9 million tons in 1911–13 to 18·6 million tons in 1918–19.

(b) Pork and lard exports more than doubled in quantity and beef exports increased sixfold.

(c) The wheat acreage increased from 48 million (19·2 m. ha) to 60 million (24 m. ha).

(d) Exports of cotton and tobacco diminished, but increased domestic demand more than compensated for this.

(e) By 1917 food prices had risen to 49 per cent above the 1913 level.

29. The Food Administration. The Food Administration (1917) was set up:

(a) to organise agricultural production;

(b) to organise the buying and selling of foodstuffs.

The rate of increase in prices was slowed down, but by October 1918 the food price index was 81 per cent above the 1917 level.

30. Farmers' gains. Considerable gains were made in real income, but gains were spread unevenly:

(a) *Agricultural wage earners* benefited only slightly; their gain was 2 per cent and farmers' gains 29 per cent.

(b) *Farmers to benefit most* were hog producers, wheat growers, cattlemen and wool producers.

(c) *Cotton growers* suffered from diminished British demand but benefited from increased domestic demand.

31. The continuation of prosperity, 1918–20. The boom conditions continued after the armistice until 1920–1. Factors accounting for this were as follows:

(a) *Russia*, Europe's main producer of cereals, was in turmoil.

(b) *Recovery from armed conflict* in Europe was slow.

(c) *Europe held only small reserve stocks* of food and had to continue buying from the U.S.A.

(d) *A world shortage of shipping* existed because of wartime losses: the U.S.A. concluded the war with a large mercantile marine.

(e) *American credit facilitated* the maintaining of a high European demand for her foodstuffs.

(f) *Inflationary tendencies* in Austria and Germany kept trade moving briskly.

32. The effects of the continued boom.

(a) *Mortgages* were taken out by many farmers either to increase land holdings or to effect improvements in anticipation of a continuance of high prosperity. (The *Federal Farm Loan Act*, 1916, facilitated this.)

(b) *Tenancy increased*, as it is apt to do in a period of increasing land values.

YEARS OF DIFFICULTY, 1920–2

33. The post-war depression. Overseas demand for American foodstuffs fell after 1919:

(a) European countries had better access to supplies from lower-cost areas of production – Australia, India, Argentina, etc. – as shipping facilities improved.

(b) Financial obligations could not be met by Europe – especially by Austria and Germany: post-war difficulties were more acute than was realised. There was a failure to export sufficient to pay for imports. Effective aid was needed, but was not forthcoming.

(c) European production of foodstuffs increased as recovery from the war was made by farmers and landowners.

34. Results of the fall in demand.

(a) Prices of American farm products tended to fall to the level of Europe's buying power. The terms of trade between American industry and agriculture changed in favour of industry in the short run.

(b) Between 1919 and the end of 1921 the net income for farmers fell by about 60 per cent *per capita* compared with a fall in non-farm incomes of about 20 per cent *per capita*.

(c) Cereal producers suffered more than most. *Per capita* consumption of dairy produce, vegetables, fruit and meat increased: that of grain did not.

(d) Farmers redoubled their efforts to compensate for price falls. Mechanisation increased, and this, in turn, helped prevent prices rising as might have occurred if production had been restricted.

(e) Land that was marginal in 1919 became sub-marginal and went out of production.

(f) Indebtedness incurred before 1920 became a much heavier burden as prices fell: many farmers had to face foreclosure of their mortgages.

AGRICULTURE AND ECONOMIC GROWTH, 1922–9

35. The boom of the 1920s. From 1922 to 1929 the American economy experienced a boom greater than any economic boom before that date.

(a) *The gross national product* increased from \$137·5 billion to \$181·8 billion (at market prices) and *per capita* income from \$55 to \$716 (at 1929 prices).

(b) *Population increased* from 110 million to 122 million.

(c) *Unemployment* in the years 1923–9 averaged only 3·3 per cent.

36. Agriculture's difficulties. Agriculture, however, continued to labour under difficulties:

(a) *Grain producers and cotton producers* suffered from a failure

of foreign markets to revive to the extent required to push prices
upwards by a significant proportion. Compared with 1910–14,
farm income remained at a low level and never in fact
recovered to the pre-war level. 1928 was the best year, when
total farm income was approximately 90 per cent of the pre-war
level.

(b) *The value of farmland* decreased, and the acreage under
crops fell by about 13 million acres (5·2 m. ha).

(c) *Per capita income* on farms fell sharply in the 1920–1
depression and did not recover to the same level, whilst non-
farm incomes did so. Average *per capita* income on farms in
1928 was about 25·6 per cent of that of non-farm *per capita*
income.

37. Reasons for difficulties.

(a) *Competition was less imperfect* than in industry. Farmers
produced more when prices fell and this had the effect of pushing
prices even lower and reducing total farm income.

(b) *The export market* for foodstuffs *did not recover* because of
competition from lower-cost areas of production – and because
of economic nationalism. Cotton markets did perk up but competition from synthetic fibres retarded a full recovery for cotton
producers.

(c) Although population increased, *its rate of growth slowed down*,
and *per capita* consumption of foodstuffs remained below the
1910–14 level.

(d) *Soil exhaustion* and soil erosion, resulting from wrong
cropping practices, made large areas infertile.

(e) *Credit facilities were inadequate* in debtor areas such as the
sharecropping areas of the South. There were a substantial number
of bank failures in agrarian regions.

38. Redeeming features.

(a) *Beef prices increased* considerably after 1925; livestock
producers enjoyed prosperous times.

(b) *Dairy farmers*, generally, experienced rising demand for
their products; truck farming was profitable.

(c) *Co-operative marketing* ventures in milk and citrus fruit were
markedly successful.

THE CRASH OF 1929 AND ECONOMIC DEPRESSION

39. Agriculture in the 1929–32 depression.

(a) In October 1929 the stock market collapsed and an economic decline set in. The gross national product fell from $181 billion to $126 billion and farmers' gross income diminished by more than 50 per cent and farm values declined by 33 per cent between 1929 and 1933.

(b) Federal assistance was rendered under the *Agricultural Marketing Act* of June 1929. The Federal Farm Board attempted to stabilise grain and cotton prices by purchase of the commodities through the Grain and Cotton Stabilisation Corporation – but with little effect. The general level of production remained too high to enable prices to be held for long.

(c) In 1930 the exports of agricultural produce were the lowest since 1915 and nearly 20 per cent below the 1928 level.

THE 1930s: AGRICULTURE AND THE NEW DEAL

Agriculture suffered from an excess capacity resulting in low farm incomes. Under the New Deal attempts were made to remedy these defects.

40. The Agricultural Adjustment Act, 1933. This Act attempted to balance production with consumption to raise prices and total farm incomes by:

(a) reduction in acreage of cotton, grain, etc.;
(b) marketing agreements;
(c) the imposition of a processing test.

Under the Act cotton and tobacco output were reduced, hogs were slaughtered and the marketing of fruit was controlled.

41. The Agricultural Adjustment Act, 1938.

(a) *Background and aims.* The 1933 Act was declared unconstitutional in 1936, and a *Soil Conservation and Domestic Allotment Act* was passed in that year to restrict crop production, but this, too, failed, making the 1938 Act necessary. Its aims were:

(i) to continue the *Soil Conservation and Domestic Allotment Act*, so as to preserve land resources by encouraging the growing of crops that would conserve soil;

(*ii*) to assist in the marketing of agricultural produce;

(*iii*) to regulate interstate and foreign trade in cotton, wheat, corn, tobacco and rice so as to attain and maintain reasonable prices for the products (parity prices) and to ensure a steady flow of supplies by the use of quotas.

(*b*) *Parity prices*. The period 1909–14 was chosen as a base period. Attempts were made to ensure that, in real terms, agricultural products had the same terms of exchange as in 1909–14.

(*c*) *Loans*. A Commodity Credit Corporation was authorised to make available loans on agricultural commodities.

42. Mortgage and credit facilities.

(*a*) *The New Deal*. The government aimed to reduce farm debts and give some security against mortgage foreclosure.

(*b*) *Federal Land Banks*. In 1933 Federal Land Banks were authorised to issue $2 billion in 4 per cent bonds, the interest to be used to re-finance farm mortgages. A Farm Credit Administration was set up.

(*c*) *Farm Mortgage Re-financing Act*, 1934. A Federal Farm Mortgage Corporation was created to aid in the re-financing of farm debts.

(*d*) *Farm Mortgages Foreclosure Act*, 1934. The authority of the Land Bank Commission was extended.

(*e*) *Frazier–Lemke Bankruptcy Act*, 1934 (*and* 1898). Farmers were given a five-year period in which to rent and re-purchase foreclosed land.

(*f*) *Mortgage Moratorium Act*, 1935. The *Frazier–Lemke Act* was declared unconstitutional in 1935: the new Act replaced it and gave farmers a three-year instead of a five-year respite.

(*g*) *Bankhead–Jones Farm Tenant Act*, 1937. This Act provided loans for farmers and share croppers who wished to own their own farms.

(*h*) *Farm Security Administration*, 1937. Attempts were made to set up an organisation to provide aid for the agricultural workers who were in need of assistance. Loans were provided and camps set up.

(*i*) *Other aids to agriculture*. These came under flood control schemes, electrification, highway schemes and reciprocal trade agreements. *Laissez-faire* was replaced by massive state intervention in the 1930s.

43. An appraisal of New Deal policies.

(a) *Attempts to eradicate surplus production* were only partially successful. The cotton, tobacco, wheat and corn acreages were reduced, but the vagaries of nature (especially the drought of 1934) were probably more effective in reducing wheat production than were New Deal policies.

(b) *National farm income* increased by 200 per cent between 1932 and 1937 but fell in 1938.

(c) *Farm price parity ratio* increased by 60 per cent between 1932 and 1937 but diminished in 1938. (In 1937 the ratio was almost at the 1929 level.)

(d) *Increased total income.* New Deal policies, therefore, succeeded in boosting farm income and in restoring farm prices to about the 1929 level, in terms of the parity index, but in 1929 American agriculture was in some difficulty. Whilst the New Deal pulled agriculture out of the bottom of the trough of depression it did not push it on to a peak of prosperity. Government payments to farmers in 1940 totalled $723 million, an indication that the crop surplus problem had not been overcome.

(e) *Mortgage foreclosure.* More than 700,000 farmers were saved from mortgage foreclosure by the *Farm Credit Act* and the *Farm Mortgage Act.*

(f) *The restriction of cotton and tobacco production* was, perhaps, too successful: restrictions had to be removed in 1935 to prevent export markets being lost. In 1938 cotton exports fell to the lowest level for sixty years because the American price had been forced above world prices. Sharecroppers and small farmers seem to have been affected more than most by crop restrictions and these were the ones whom the New Deal was supposed to assist.

(g) *Sow slaughter* reduced pig output and so did the increases in the prices of feeding stuffs: the total reduction was greater than intended.

AGRICULTURE, 1939-50

44. The Second World War and afterwards.

(a) *Lend-lease* stimulated American agriculture: farm income began to rise.

(b) *Britain's imports* from North America trebled in value between 1939 and 1944. The U.S.A., once more, became the safest, most convenient market from which products could be purchased, and American loans facilitated the expansion of British imports.

(c) *Pressure on labour supplies* helped increase the drift from the land. The agricultural labour force declined from 9·54 million in 1940 to 7·5 million in 1950, but production increased, aided by much greater mechanisation. The *Federal Reserve Bulletin* (August 1960) shows that the value of machinery and motor vehicles (at 1940 prices) more than doubled between 1940 and 1960.

(d) *Agricultural productive capacity* was stretched during the war, and demand from abroad after the war was maintained by the Marshall Aid Programme. Prices remained high: total farm income (at constant prices) nearly trebled between 1940 and 1944.

PROGRESS TEST 6

1. Discuss the effects of the extension of the agricultural frontier after 1860. **(1, 5, 15)**

2. What impact did the Civil War have on southern agriculture? **(14)**

3. Why was there unrest among American farmers in the last quarter of the nineteenth century? **(19)**

4. Outline the main farmers' movements. **(20–25)**

5. What geographical shifts in production occurred before the First World War? **(26)**

6. Explain the effects of the First World War on American farming. **(27–32)**

7. Why was American agriculture in some difficulty after 1920? **(33–38)**

8. Discuss the *Agricultural Adjustment Acts*. **(40, 41)**

9. Discuss the effects of the Second World War on agriculture in the U.S.A. **(44)**

PART FOUR

INDUSTRIAL DEVELOPMENT

CHAPTER VII

THE GROWTH OF INDUSTRIALISM BEFORE 1914:
COMPARATIVE ASPECTS

1. The change from agrarian to industrial economies. Key periods in the development of industry were the first half of the nineteenth century in Britain and the second half of the century in the U.S.A. Rostow dates the stages of take-off into sustained economic growth as roughly from 1783 to 1802 in Britain and the 1850s in the U.S.A., but the recent trend is to view with scepticism attempts to pinpoint fundamental industrial changes to one sector and one short period of time and to look at aggregate effects over long periods.

(*a*) Relatively, *agriculture declined* in importance in both countries between 1850 and 1914. In terms of contribution to the national product:

(*i*) Britain's agriculture was outpaced by industry by 1830; and
(*ii*) American agriculture was overtaken by industry by 1890.

(*b*) *Society was transformed* in both countries by:

(*i*) the growth of urban areas;
(*ii*) the development of mechanical transportation, as the steam engine was improved and put into general use;
(*iii*) the use of powered apparatus in industry, which transformed the manufacture of textiles and the output of iron and steel;
(*iv*) the population explosion accompanied by an increase in agricultural output;
(*v*) the discovery and exploitation of vast reserves of natural resources facilitated by technological progress in metallurgy, machine tools, chemistry, transport and business organisation.

2. American industrial acceleration.

(a) Initially the U.S.A. leaned on European technology and financial facilities:

  (i) Great use was made of British expertise in textiles, iron and steel production, coal-mining, etc.

  (ii) By making use of European capital to finance transport development and mining, the U.S.A. released capital to finance other industrial growth. The proportion of national income invested to bring about industrialisation was probably less in the U.S.A. than in Britain because of the availability of foreign capital for the former's initial industrialisation: the latter relied primarily on her own capital.

(b) The U.S.A. surged ahead of Britain in industrial output in the second half of the nineteenth century and became the world's leading industrial nation. By 1900 American steel production was double that of Britain's, and productivity of labour in the industry was about double that of British labour.

(c) By 1914 the U.S.A. possessed about 40 per cent of the world's manufacturing capacity.

3. Industrial growth and trade.

(a) *British industrialism* was closely linked with the international economy as well as with domestic markets. She relied increasingly on imports of raw materials and foodstuffs and exported large quantities of manufactures.

(b) *Industrialists in the U.S.A.* were more introspective because of the rapid and sustained growth of the domestic market and because of the huge domestic reserves of raw material resources and agricultural land, but the U.S.A. became important as an exporter of manufactures by 1914.

(c) *Different tariff policies* were adopted. Britain remained a "free trade" nation, whilst the U.S.A. adopted protective tariffs for her industry.

4. Investment banking.

(a) Bankers were closely concerned with industrial growth and especially industrial combination between 1865 and 1914 in the U.S.A.; long-term capital was provided.

(b) In Britain the role of the banker was confined to the provision of working capital: there was not a banker-domination of industry as in the U.S.A.

5. The U.S.A. as industrial pioneer. Just as Britain had been the main pioneer in the early nineteenth century, so the U.S.A. became a pioneer in the industrial field after 1840:

(a) Mass production of products such as sewing machines, typewriters, revolvers, motor cars, etc., which depended upon a huge domestic market and sophisticated machine tools, was considerable.

(b) The development of scientific management, which laid the foundations of large-scale industrial firms, occurred.

(c) A harnessing of science for industrial progress, with emphasis on research and education in science, was a key factor in technological progress.

PROGRESS TEST 7

1. What were the reasons for the changes from agrarian to industrial economies? **(1)**
2. In what ways did Europe facilitate American industrial growth? **(2)**
3. How important were investment banks in the growth of industry? **(4)**
4. To what extent was the U.S.A. an industrial pioneer? **(5)**

CHAPTER VIII

INDUSTRY IN BRITAIN

INDUSTRY BEFORE 1850

1. The role of the cotton industry in economic development. By 1850 the industrial revolution was in full sail:

(a) A railway network was spread over the nation.

(b) Investment in fixed plant was relatively large and was increasing.

(c) The production of iron and coal was increasing rapidly.

(d) Factory-type production was becoming more and more significant.

(e) Urban areas had expanded enormously.

The older school of economic historians asserted that the cotton industry was the key factor in this process of industrialisation and more rapid economic growth before 1850, because of the multiplier effect of initial developments which induced fundamental changes in the structure of the economy. This view is challenged by recent researchers.

2. The aggregate view. The hypotheses about stages in economic growth and the role of the cotton industry in Britain have been subjected to a critical reappraisal on several grounds:

(a) The interrelationship of the cotton industry with other industries was not such that changes in the industry would bring about fundamental changes in other sectors.

(b) Economic development is a result of a number of changes and forces working out over a relatively long period of time. Massive structural change in the economy of Britain before 1850 is more likely to have been set in motion by investment in railways between 1830 and 1850 than by investment in the cotton industry between 1783 and 1802 (the so-called "take-off" stage), but one factor should not be considered in isolation from others, and a period of more than twenty years is probably necessary for fundamental structural changes to take full effect in the economy.

(c) Agriculture was significant in the transformation of the economy:

(i) Investment in agriculture was high enough to bring about changes in production that enabled the industry to provide foodstuffs for a rapidly increasing population, at a time when large food surplus areas abroad did not exist and long-distance transport facilities were relatively poor and expensive to use.

(ii) In 1815 Colqhoun estimated that 64 per cent of the nation's capital was invested in land, farm implements and farm buildings and only 19 per cent in industry, transport and commerce.

(iii) Changes in agriculture were part of a whole series of changes in the development of the economy. The industry made important, significant contributions to the national income not only from 1783 to 1802 but for many years afterwards. In 1851, together with forestry and fishing, it still contributed more than 20 per cent of the national income and employed more than 30 per cent of the labour force. (The value of the net output of all textile industries together, in 1850, was only 10 per cent of the national income and that of cotton alone about 4 per cent.)

(d) Other factors of crucial importance were developments in metallurgy, transport, foreign trade, the adoption of steam engines, etc., without which industrialisation, as it occurred, could not have taken place.

THE SPREAD OF INDUSTRIALISM

3. The extent of industrialisation in 1850.

(a) *The cotton industry* was largely mechanised. Steam power was being used on an increasing scale. Handloom weavers constituted only one-seventh of the labour force; fifteen years earlier they had represented 50 per cent of the labour force. There was heavy investment in power looms by entrepreneurs between 1830 and 1850.

(b) *Developments in steam-engine technology* were rapid after 1830 because of the work of people such as Fairbairn, Hetherington, Nasmyth, etc. The use of steam hammers made the forging of large shafts feasible and improvements in metallurgy facilitated the development of high-pressure steam boilers, making bigger engines possible.

(c) *A machine-tool industry* was rapidly developing, based upon the work of inventors and innovators such as Bramah, Maudslay, Brunel, Roberts and Whitworth. This was a key point in further industrialisation; it facilitated progression based upon precision and scientific methods.

(d) In *the iron industry* the quantity and quality of output were improved by the discoveries of Nielson and Homfray, and by the use of Nasmyth's steam hammer. Large-scale units in the industry began to develop: the Dowlais Iron Works had eighteen furnaces, each producing more than 3 tons of iron per week; at Cyfarthfa Iron Works there were eleven furnaces. In Scotland the average total output of furnaces was 120 tons per week. Estimates indicate that Britain was producing about half of the world's pig iron in 1849; the bulk of the British "make" was produced in three areas – Scotland, Staffordshire and South Wales.

(e) *Coal output*, by 1850, exceeded 49 million tons per year. Technological developments made it possible to work seams at greater depths. Coal exports exceeded 3 million tons in 1851, having approximately doubled over the previous decade.

LIMITED LIABILITY

4. Company law in 1850. Limited liability for investors and partners in joint-stock companies could be obtained only under a Royal Charter or by means of a private Act of Parliament. Considerable expense was involved in securing the facility of limited liability.

(a) *Arguments in favour of* general limited liability for industrial and commercial companies were as follows:

(i) Able but poor men found it difficult to raise capital for industrial and commercial projects because of existing company law.

(ii) Small savers with capital to spare were precluded from investing in all but the smallest of industrial and commercial concerns.

(iii) Richer individuals with large amounts of surplus capital were reluctant to invest in companies whose fortunes they could not control whilst unlimited liability was required. Broadbridge illustrates this in the case of Lancashire mill owners between 1845 and 1855. Surplus capital existed but investment was not forthcoming because of the fears of the consequences of unlimited liability should failure occur.

(*iv*) The *Health of Cities Report* (1845) noted that public utility companies were hindered in development. Limited liability would have facilitated the obtaining of capital for such socially useful projects. The Christian Socialists urged reform of company law to enable municipal enterprises to make more rapid progress.

(*v*) Railway companies, etc., secured limited liability and illustrated the potential success of enterprises able to harness large amounts of capital under the umbrella of limited liability. Industry, as a whole, desired the same facilities.

(*vi*) The existence of limited liability abroad was thought to be a cause of the export of British capital at the expense of domestic investment.

(*b*) *Arguments against* general limited liability were as follows:

(*i*) Capital in Britain was abundant: limited liability was unnecessary.

(*ii*) General limited liability would encourage speculative investment and would damage foreign confidence in British firms.

(*iii*) Controllers of limited liability firms would not display the initiative and efficiency without the spectre of the consequences of unlimited liability.

5. The development of general limited liability.

(*a*) The *Bubble Act* of 1720 was repealed by the 1825 Act. Liability could be prescribed; it was not limited at common law and had to be written into contracts individually unless the company concerned was incorporated under Charter or Act as a company with limited liability.

(*b*) The *Trading Companies Act*, 1834, gave unincorporated companies the right to litigate through their chief officers but did nothing towards facilitating incorporation with limited liability.

(*c*) In 1837 the *Charted Companies Act* (sometimes known as the "*Letters Patent*" *Act*) gave the President of the Board of Trade the opportunity to widen the avenue to limited liability by granting charters by "letters patent." Use was made of the procedure for mining companies, railway companies, insurance companies and literary societies, but it appears that the only manufacturing company to submit a successful application was the British Plate Glass Company, and, in every case, expenses continued to be high.

(*d*) The *Registration Act* of 1844 defined joint-stock companies and compelled them to register at the Joint Stock Companies Registration Office disclosing the names of promoters and the objects of the companies. Full registration could be achieved only on the execution and registration of a deed of settlement signed by at least 25 per cent of the stockholders possessing together at least 25 per cent of the capital. The registrar had power to reject deeds of settlement.

Although companies obtained legal recognition, liability of members remained unlimited and continued for three years after shares were sold.

(e) The *Report on Savings* (1850) and the *Report on Partnership* (1851) stressed the hindrance to economic development caused by company law.

(f) The *Limited Liability Act*, 1855, enabled companies to obtain limited liability by registering under the 1844 Act, provided companies had twenty-five shareholders and 15 per cent of the capital was paid up.

6. The extension of limited liability.

(a) The *Joint Stock Companies Act*, 1856, repealed the 1855 Act:

(i) Any seven or more persons could register a memorandum of association and form a limited company.

(ii) A model set of company rules was drawn up which companies could use if they wished.

(b) The *Limited Liability Act*, 1858, extended the facility to banking companies.

(c) The *Companies Act*, 1862, consolidated previous legislation and permitted insurance companies to be formed as limited liability companies.

7. The effect of limited liability on industrial organisation.

(a) Limited liability as a general right caused no sudden revolution in the organisation of firms in industry. The change-over to incorporation with limited liability was gradual: it is estimated that in 1914 there were about 62,000 firms incorporated under the 1862 Act and in 1939 approximately 162,000.

(b) Between 1856 and 1883 more than 20,000 companies were registered, but many failed. Early use was made of the limited liability form of organisation by firms in engineering, coal-mining, railway development, shipbuilding and iron and steel production. In these areas of production large capital outlays were necessary.

(c) In the 1870s there was an extension of company organisation (with limited liability) among Lancashire cotton firms, and by 1914 such firms dominated the industry.

(d) The adoption of limited liability and incorporation as a public company enabled firms to increase the size of the pool of capital on which they could draw and facilitated an increase in the size of operations. This had certain effects:

(i) It led to greater stability in the long run.
(ii) Integration was facilitated.
(iii) A separation of ownership from management gradually developed: professional managers began to run industry.

THE GROWTH IN SIZE OF FIRMS BEFORE 1914: COMBINATION OF UNITS

8. Types of combination. More intensive competition, increased capitalisation of firms and falling prices in the later part of the nineteenth century led to greater combination among firms in industry. Firms combined:

(a) by merging one with another by agreement;
(b) by "take-overs," one firm acquiring control of other firms;
(c) more loosely, through trade associations, or by use of cartel arrangements, some of which were international in character.

9. Integration of industrial firms.

(a) Integration of firms was both *vertical* and *horizontal* and occurred in various industries in the 1890s and early years of the twentieth century – iron and steel, chemicals, wallpaper, tobacco, soap, cement, etc.

(b) Compared with business consolidation in the U.S.A. integration of firms in Britain was on a relatively small scale. This was due to various factors:

(i) Bankers were closely concerned with American industrial firms and provided much long-term capital. The banks were prominent in instigating and controlling the consolidation of industrial units. In Britain the role of banks in industry was less dominant. Working capital was provided but there was no emphasis on long-term investment in industry.

(ii) Protective tariffs sheltered American industry and provided a better climate for the build-up of monopolistic units than did the free-trade policy of Britain.

(*iii*) The scale of enterprises frequently tended to be greater in the U.S.A. because (1) the domestic market was larger than Britain's; (2) *per capita* income was higher; (3) the market was more homogeneous in taste and thereby encouraged large-scale mass production.

(*iv*) Relatively lax laws in states such as New Jersey, West Virginia and Massachusetts facilitated the build-up of industrial empires through the holding company device. There was much more rapid general acceptance of incorporation with limited liability as a basis of company organisation in the U.S.A. than in Britain.

10. Vertical integration. Integration of firms reaching backwards towards raw materials and forwards towards markets took place between 1894 and 1902, particularly in the iron and steel industry. The diagram illustrates this:

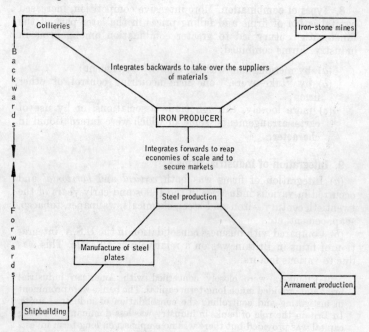

11. Horizontal integration.

(*a*) This was on a wider scale than vertical integration, and occurred among salt producers, chemical companies, cement producers, wallpaper manufacturers, railway companies, coal-mining companies, etc. For example:

| COAL PRODUCER | + | COAL PRODUCER | + | COAL PRODUCER |

(b) Some combinations were a complex of *vertical, horizontal* and "*diagonal*" integrations (especially after the First World War).

(c) Monopolistic situations developed in salt, wallpaper, sewing cottons, tobacco and cement industries because of combinations of firms.

12. Trade associations. By 1914 trade associations existed in most industries. These associations were formed by firms:

(a) to limit competition;
(b) to share out markets;
(c) to have common policies in dealing with labour.

To some extent trade associations provided an alternative to amalgamation and integration: they provided a means through which small firms could follow agreed common policies.

MONOPOLISTIC COMPETITION: COMBINATION OF FIRMS AFTER 1914

13. The old export industries. In the staple industries some reorganisation was necessary because of changes in the international and domestic market situations: markets abroad were not steadily expanding ones in the inter-war years. Difficulties were encountered in retaining markets in coal, cotton, iron and steel and shipbuilding.

(a) *The coal industry.* Amalgamation of collieries took place between 1920 and 1930, and cartel schemes were evolved in Scotland, South Wales and the Midlands. The *Coal Mines Act* of 1930 introduced a compulsory cartel scheme, and set up a Coal Mines Reorganisation Commission, but this was not very effective. In 1938 a further Commission was set up to amalgamate units but war halted progress. Nationalisation of the industry was effected in 1948.

(b) *Textiles:*

(i) The cotton industry was hard hit by loss of export markets and required reorganising into larger units to reap the cost benefits from large-scale production, but amalgamations were effected only slowly. In 1929 the Lancashire Cotton Corporation, the Combined Egyptian Mills Ltd. and the Quilt Manufacturers' Association were set up, but in 1939 more than 65 per cent of the industry remained in small units.

(*ii*) The woollen and worsted industry remained largely composed of small units and productivity was well below that achieved in the American industry.

(*c*) *Shipbuilding*. A world surplus capacity in ships and shipbuilding induced mergers in the 1920s in the Belfast, Clydeside and Tyneside areas, and the depression of 1929–33 caused the industry to form the National Shipbuilders Security Ltd. to rationalise production. Insufficient was done to build up large efficient units even in the years 1945–50.

(*d*) *Iron and steel*. Amalgamations between 1918 and 1932 resulted in ten groups achieving control of almost 50 per cent of the pig-iron-producing capacity of Britain and 60 per cent of the steel-producing capacity. In 1934 the British Iron and Steel Federation was formed, but amalgamation did not have a very significant effect on the pitch of efficiency. In 1949–51 the industry was nationalised – only to be denationalised after the Conservatives took office.

14. Expanding industries.

(*a*) *Electricity supply*. Rationalisation of supply was imposed by the state in 1926 when the Central Electricity Board was set up, authorised to concentrate electricity generation at selected stations to supply a national grid. A large number of small stations were closed down. Output more than quadrupled between 1925 and 1939 and prices fell.

(*b*) *Electrical engineering*. Amalgamations among firms in this expanding industry produced two large combinations, E.M.I. and the Electric Lamp Manufacturers' Association, which were international in character. Other large-scale groups were formed, but a great number of small firms survived.

(*c*) *Motor vehicle industry*. In the early 1920s the number of producers was ninety-six: in 1939 there were twenty and of these six firms dominated the market. After the Second World War further amalgamations followed so that in 1969 four groups – British Leyland, General Motors, Ford and the Rootes–Chrysler combine – produced almost all the vehicles in the industry.

(d) *The chemical industry.* By the end of the First World War the industry was dominated by four groups – United Alkali, Brunner Mond, British Dyestuffs Corporation and Nobel Industries. In 1926 these four groups combined to form Imperial Chemical Industries, which produced a large variety of products and was large enough to compete with foreign groups on equal terms.

15. Cartels.

(a) The state encouraged the development of cartels in industry in the 1930s – in coal, cotton, fishing, iron and steel, shipbuilding and agriculture. The aim was to maintain prices at a desired level by restricting competition and by controlling output.

(b) International cartels were developed to a greater extent than before 1914:

(i) Imperial Chemical Industries had agreements with Du Pont of the U.S.A. and I.G. Farbenindustrie of Germany; it had interests in companies in Canada, Argentina, Australia, New Zealand, Africa, etc.

(ii) Courtaulds, besides having its own companies in the U.S.A., Canada, France, etc., had linkages with firms in Germany, Italy, Switzerland, Japan, Argentina, etc.

(iii) Lever Brothers linked with Dutch margarine combines to form Unilever in 1929; the combined group had world-wide involvements.

16. Trade associations.

(a) The government fostered the growth of trade associations during the First World War because of administrative convenience in holding consultations with business and industrial interests represented by associations.

(b) The associations differed from their pre-1914 counterparts. They became separated from the employers' associations dealing with labour and confined their attention to marketing and supply.

(c) Resale price maintenance was developed, supported by "blacklists" and quotas during the 1930s.

THE EFFECTS OF SIZE AND RESTRICTION OF COMPETITION

17. The effects of the combination of industrial firms.

(a) The growth in the size of firms was accompanied by a growth in the use of the joint-stock form of organisation and increased the separation of ownership of capital from control. A burgeoning of management oligarchies occurred: the ordinary shareholder had little or no influence over policies.

(b) *Holding companies* became a common means of controlling large groups. Interlocking directorates were common. Concentration of control of industry went well beyond what was displayed on the surface in amalgamations and take-overs. More complex problems in business and industrial administration were pushed to the surface.

(c) *A greater proportion* of the labour force was employed by large firms, and a greater stability of employment resulted from increases in size.

(d) There was *a restriction of competition* and the creation of monopolistic markets, which enabled prices to be controlled more easily by the large organisations. Competition tended to move from price to service.

(e) Larger units offered *better facilities* for specialisation of function and for research, whilst financial control led to a rationalisation of resources.

THE COAL INDUSTRY

18. Production.

(a) *Production grew steadily to* 1913. In 1854 the U.K. output was 65 million tons; in 1913 it was 287 million tons. This was accompanied by a growth in exports from about 7 per cent of total output in 1850 to 33 per cent in 1913.

(b) *The value of output increased* but there were considerable fluctuations from year to year because of changes in demand and a relatively inelastic supply situation in the short term. For example, in 1900 output was 225 million tons and total value £121·7 million; in 1902 the respective totals were 227 million tons and £93·5 million. The total value of output for the U.K. approximately increased sevenfold between 1854–9 and 1909–13 and in 1907 mining was the most important industry in terms of value of net output.

(c) *The main coalfields* were located in the following regions (in order of total output in 1913):

    (i) South Wales   (3)
    (ii) North-east England   (1)
    (iii) Yorkshire   (6)

(iv) Scotland   (5)
(v) Midlands   (7)
(vi) Lancashire   (2)
(vii) Staffordshire   (4)

Figures to the right indicate order in terms of total output in 1854.

(d) *Production displayed a general trend downwards* from 1913 although the trend was reversed in 1922–3, 1929 and 1939–40. In 1945 the major producing areas were the Midlands, Yorkshire and the North-east. Between 1913 and 1945 production declined by approximately 60 per cent in South Wales and Lancashire and 50 per cent in Scotland. Total output in 1945 was down to 175 million tons; it increased to 204 million tons by 1950 (including opencast output of 12 million tons).

19. Markets abroad.

(a) Exports *increased enormously* between 1850 and 1913 so that in the latter year 98 million tons (including foreign bunker coal) out of the total output of 287 million tons went abroad and accounted for approximately 10 per cent of the total value of exports. H. S. Jevons estimated that British trade in coal in 1913 was more than double that of the rest of the world.

Exports increased because of the following factors:

(i) The needs of steam-powered ships and the world-wide bunkering services that were built up.

(ii) The increasing use of coal abroad in countries where known reserves were difficult to work or where coal-mining had not been sufficiently developed to satisfy demand.

(iii) The existence of a large British mercantile marine able to carry coal on outward journeys instead of voyaging in ballast.

(b) During and after the First World War coal exports *tended to decline* in the long run. There were years when the trend was reversed; *e.g.* in 1923 exports were boosted because of strikes in American coalfields and because of the disruption of coal production in the Ruhr, following the occupation by France.

The decline in exports was due to the following factors:

(i) Relatively high costs of production in Britain's coalfields. The industry was slow to mechanise and productivity was inferior to that in Germany, Poland, Holland and the U.S.A. by the 1930s. Seams were becoming more costly to exploit and the industry did not rationalise itself sufficiently; units were too small to reap economies of scale.

(ii) The restoration of the pound sterling to the gold standard at a high parity.

(*iii*) An increasing use of alternative fuels.

(*iv*) Economies achieved in the use of coal by the increasing adoption of more efficient apparatus which conserved fuel.

20. The home market. In 1850 the consumption of coal in the home was as important a source of demand as the manufacturing and iron and steel markets, but by 1869 the manufacturing and iron and steel sectors provided the major markets. Between 1850 and 1913 the national demand for coal more than quadrupled, but the increase in demand between 1900 and 1913 was very small – about 8 million tons. From 1900 to 1930 demand in the home market changed very little, for the following reasons:

(*a*) Alternative fuels became more economically attractive.

(*b*) There was an increase in consumption of electricity, which economised on the use of coal.

(*c*) Domestic fireplaces changed in design and became more economical in the use of fuel, and equipment used in the iron and steel, gas, electricity and shipbuilding industries also improved in design and used considerably less fuel per unit of output.

(*d*) The use of steam engines on railways began to diminish. Diesel engines began to replace them, and there was a transference of traffic from rail to road which affected the demand for coal. The economies counteracted the effect of any increase in demand caused by an increase in the number of users of coal as a fuel.

21. The labour force. Between 1851 and 1914, the labour force of the industry increased nearly fourfold – from 300,000 to 1,134,000. There was a decline in the labour force in 1915, but it then increased to a peak of 1,248,000 in 1920, after which the long-term trend was one of declining numbers, and by 1933 the labour force was less than 800,000. It increased during the Second World War and in the immediate post-war years when there was a dearth of coal, but increasing use of alternative fuels caused a decline in demand for coal in the late 1950s, and this was accompanied by a diminution of the labour force so that in 1966 less than 600,000 were employed in mining.

22. Unemployment. Large-scale unemployment was a severe problem in the industry from 1924 to 1939. From 1927 to 1936 more than 19 per cent of the miners were out of work and the proportion was in excess of 27 per cent between 1931 and 1935.

The causes were a tendency for exports to decline and a failure of home demand to increase to any appreciable extent. Economic depression in the western world was a key contributory cause of the massive unemployment between 1931 and 1934.

23. Industrial relations, 1850–1914.

(a) Despite the collapse of the Miners' Association in 1847, county mining unions spread in the next two decades, especially in Scotland and the north-east of England, and a National Miners' Union was formed in 1858. The *Coal Mines Regulations Act*, 1860, compelled employers to employ checkweighmen appointed by the miners and the checkweighmen became the focus for union activities at collieries. County unions continued to be of key significance and in 1869 the Amalgamated Association of Miners was formed in Lancashire. The two national unions grew rapidly and by 1873 had a combined membership of 200,000, but the latter had collapsed by 1880 because of adverse economic circumstances.

(b) There was a growth of unions in the industry in the late 1880s and the Miners' Federation of Great Britain was inaugurated in 1888. Sliding scales adopted by Conciliation Boards were opposed because no minimum was fixed and concessions were secured from employers in relation to sliding scales.

(c) By 1906 costs in the industry were at a level that caused difficulty in wage negotiations and were due to:

(i) falling productivity;
(ii) the *Coal Mines Act* of 1896, workmen's compensation, a coal export tax (1901), timbering and haulage regulations and the *Coal Mines Regulations (Eight Hours) Act*, 1908;
(iii) less accessible seams.

(d) Conflict between workers and employers followed and a national strike occurred in 1912.

24. The General Strike, 1926.
Industrial relations were exacerbated in 1921 after the decontrol of the industry and a bitter strike followed. In 1926 the employers cut wages, increased the working day from seven to eight hours, and substituted regional for national wage agreements. A general strike followed a lock-out of the miners, who stayed out for seven months. Relationships between employers and workers were embittered for many years to come.

25. The National Union of Mineworkers.
In 1944–5 the Miners' Federation became the National Union of Mineworkers. The

N.U.M. was made up of forty-one constituent bodies which were reorganised into twenty-one area organisations. The miners remained somewhat militant even after nationalisation of the industry in 1948.

26. Entrepreneurs and organisation. In the nineteenth century the size of units in coal-mining varied enormously – from large undertakings with considerable fixed capital investment, in which large landowners such as the Duke of Buccleuch were involved, to small undertakings worked by the entrepreneur himself and a few employees.

(a) *Integration.* Demand for coal and iron tended to be joint demand and iron producers integrated vertically in a backwards direction to ensure that supplies of fuel were forthcoming at required times and at the right price. This integration process led to the build-up of large firms which dominated certain areas; *e.g.* the Crawshay, Guest, Powell Duffryn and Cambrian firms dominated South Wales, the Pearce family rose to a position of dominance in the north-east and the Bairds and Dixons became dominant in Scottish coalfields. Integration of units in the post-1873 period meant that employers and employees lost personal contact with each other and mutual sympathy and understanding diminished. Much of the bitterness and tension manifest in industrial relations in the industry in the twentieth century was a result of conflict in the second half of the nineteenth century. However, increasing demand for coal before 1914 encouraged the rise of a large number of separate coal-producing units. Even in 1914 there were more than 2,700 mines with an average output of less than 100,000 tons per year. A substantial number of collieries each employed less than a hundred men.

(b) *Nationalisation.* Nationalisation was advocated by miners and others before 1919 on economic and political grounds. In 1919 the Sankey Commission appeared to favour such a measure but not until 1948 was nationalisation effected. The industry failed to rationalise and modernise itself to a sufficient extent in the inter-war period. There was a lack of investment and this caused difficulties in maintaining supplies of coal at the required level during the Second World War and in the immediate post-war years.

THE COTTON INDUSTRY

27. Production, 1850–1950.

(a) *From 1850 to 1914* production of cotton goods increased so that, whereas imports of raw cotton retained in Britain were 707 million lb (320·7 m. kg) in 1850, they were approximately 2,000 million lb (907 m. kg) in 1913. The growth in production was not constant: setbacks occurred in 1860–4 (because of the American Civil War and the reduction of American exports of raw cotton), 1875–9 and 1885–9. From 1900 to 1913 there was a greater concentration on the production of finer yarns and cloths, and the value of total output increased proportionately more than physical output.

(b) *In the First World War* production decreased because of demands on shipping space and the consequent reduction in raw-cotton imports, and there was a shortage of manpower. A brief boom in demand was experienced in 1919 and 1920 but production then declined more or less continuously until 1945. In 1951 yarn output was still only four-fifths of the 1939 level and piece-goods output about two-thirds of the 1939 level of production.

(c) *By 1939* the cotton industry was no longer predominantly an export industry. Declining output was due largely to a loss of export markets. Home demand in the 1930s exceeded the 1913 level.

28. Markets abroad, 1850–1914.

(a) *Yarn exports* went mainly to Europe in 1820–40, but by 1880 50 per cent went to India, the East Indies, China and Japan. After 1880–1900 there was a decline in demand for British yarns in eastern markets.

(b) *Piece goods*, up to 1850, were exported mainly to Europe and North and South America but by 1911 Europe was taking only 6 per cent and the U.S.A. only 1 per cent of British exports of piece goods. South and Central America became more important markets, but the chief markets were the ones in the East – China, Japan, etc.

(c) In *cotton goods*, Britain's share of the world export trade changed very little during the period 1870–1914, and in absolute terms trade increased by approximately 80 per cent in value from 1866–70 to 1910–14. In volume the increase was greater than this.

29. The loss of export markets after 1914. The First World War hastened on a process that was apparent to some contemporary

observers – the increase in production of cotton textiles in what were formerly British market areas for cotton goods, *e.g.* Japan and India. A brief boom in demand for British exports occurred between 1918 and 1920, but over the period 1913–45 exports declined. By 1937 yarn exports were running at about two-thirds of the 1913 level and piece goods at about a third.

Causes of the loss of markets were as follows:

(*a*) In the First World War shipping space for textile exports and raw-cotton imports was restricted. In 1917 the Cotton Control Board rationed the supply of raw cotton and cut back production of cotton cloth intended for markets in India, Japan, the Near East and the Balkans. This presented producers in India and Japan with opportunities to seize markets deprived of imports from Britain.

(*b*) Tariff barriers were erected in the inter-war years in the U.S.A. (by the Fordney–McCumber Tariff of 1922 and the Hawley–Smoot Tariff of 1930) and Brazil.

(*c*) France and Germany imposed import quotas.

(*d*) The collapse of the American economy after 1929 hastened the growth of economic nationalism and moved world trade from a multilateral to a bilateral basis with a subsequent reduction in the volume of world trade in the 1930s compared with the 1920s.

(*e*) Britain's cotton industry remained relatively a high-cost one compared with the Japanese and Indian industries. Rationalisation and modernisation were required.

(*f*) In 1925 the pound sterling was restored to the gold standard at a parity that Keynes estimated was 10 per cent too high.

(*g*) The Second World War caused a further loss of markets, especially to American producers.

(*h*) Synthetic fibres became more competitive and possessed certain advantages compared with cotton fibres.

(*i*) After the Second World War competition in markets gradually hotted up in the 1950s. Lancashire producers found their home market invaded by Japanese, Hong Kong and Portuguese textiles.

30. Employment and the labour force.

(*a*) *Employment.* Factory production became more important than handloom work, in respect to the numbers employed, by

INDUSTRIAL DEVELOPMENT 87

1834, and by 1862 the number of handloom workers had diminished to about 3,000, compared with the 450,000 working in factories. The peak year for total employment was 1911, when some 646,000 were employed. A decline in numbers employed then set in; in 1930 less than 500,000 worked in the industry, by 1939 less than 400,000, and in 1945 the number was less than 250,000.

(b) *The labour force.* Female operatives made up the majority of the labour force throughout the period 1850–1950, and in the nineteenth century children under 14 formed an important part of the working force in the industry. In 1874 nearly 70,000 children under 13 were employed; in 1907 the number of children under 14 who worked part-time was about 20,000.

31. Entrepreneurs, organisation and technology.

(a) Two types of producer predominated in the early nineteenth century:

(i) Large-scale producers such as Horrocks, the Strutts, the Arkwrights, etc., who employed large numbers of workers in factories – between 200 and 300 in each factory.

(ii) Small-scale producers who had relatively small workshops, or who still operated on the domestic system.

(b) It was the large-scale producers who benefited most from development in steam engines and automatic machines in combing, spinning and doubling, and weaving, for the following reasons:

(i) They either had adequate capital reserves to utilise the latest developments or had access to capital.

(ii) They produced on a scale sufficient to make the adoption of new improvements worth while economically.

(c) In the period 1850–70 great strides were made in applying steam power to production, but otherwise changes in technology were more in the nature of gradual modification of existing machines in the industry. Mules and power looms were improved, but the industry was slow to adopt ring spinning, automatic looms and electricity as a motive power. Lancashire suffered from complacency in the early twentieth century: the effects of this attitude were shown after 1920.

(d) There were about 1,600 firms in the industry in 1880–4, the majority being family partnerships or single-proprietor firms:

(i) Joint-stock companies with limited liability in the industry were found mainly in spinning. Approximately 25 per cent of Lancashire cotton-spinning firms were so organised in 1885. By 1907 cotton spinning was mainly in the hands of such companies, who produced standardised goods for bulk sale. In 1898 thirty-one firms grouped together to form the Fine Cotton Spinners' and Doublers' Association Ltd., but many small firms remained in existence in 1913.

(*ii*) Weaving was largely controlled by private firms and private companies, even in 1919.

(*iii*) Approximately a third of total capacity in finishing was in the hands of public companies.

(*e*) After 1919–20 the reduction in exports left the industry with surplus capacity, and there was an urgent need for reorganisation, but most firms were small and agreement proved difficult to achieve.

(*f*) Efforts were made to improve efficiency by amalgamation:

(*i*) 1919. The Amalgamated Cotton Mill Trust Ltd. was formed.

(*ii*) 1920. Crosses and Heatons amalgamated.

(*iii*) 1917–20. Joshua Hoyle & Sons took over other firms.

(*g*) Efforts were made to restrict output:

(*i*) 1927. The Cotton Yarn Association was set up to control output (this was a failure).

(*ii*) 1929. The Lancashire Cotton Corporation was established to rationalise the American section of cotton production. Combined Egyptian Mills Ltd. was set up to rationalise the Egyptian section. Both these schemes had some success.

(*iii*) 1936. The Spindle Board was set up to acquire and scrap spindles.

(*iv*) 1939. The Cotton Industry Board was established to fix minimum prices.

THE IRON AND STEEL INDUSTRY

32. Production, 1850–1950.

(*a*) Technological change influenced the *nature of production* between 1850 and 1950. The chief end-product in the period 1850–80 was iron, but from 1870 onwards steel became more and more important and by 1920 exceeded iron in volume and value of output. The Bessemer, Siemens–Martin and Gilchrist–Thomas processes (*see* **33**) made this possible by enabling pig iron to be converted into steel cheaply. Production of steel increased from 1 million tons in 1879 to 5 million in 1903 and to 9·7 million in 1917; it then fell to 3·7 million in 1921. In the following years output fluctuated tremendously from 8·5 million tons in 1923 to

3·6 million in the year of the General Strike (1926), and then to the peak of the 1920s of 9·6 million in 1929. In the great depression which followed output fell but in 1939 it stood at a new record level of 13 million tons, which was not exceeded until 1948.

(b) In 1870 Britain was the premier producer of iron and steel: she produced about half the world's output of both products. In 1901 both the *U.S.A.* and *Germany* had overtaken Britain in iron and steel production. In that year Britain's share of world output was 20 per cent of iron and less than 20 per cent of steel, but the decline was a relative not an absolute one.

The reasons for the greater rate of growth of output in the U.S.A. and Germany were as follows:

(i) Their more abundant natural resources, made more significant by the Gilchrist–Thomas process of steel making.

(ii) The imposition of tariff barriers which afforded protection and thereby made the marketing of Britain's iron and steel products more difficult in the U.S.A. and Germany.

(iii) The existence of larger domestic markets, especially in the U.S.A., which stimulated the growth of large-scale units able to reap the economies of scale. (In the U.S.A. the population exceeded 90 million by 1911 and *per capita* income was higher than in Britain.)

(iv) A greater willingness to adopt newer, more efficient methods of production with an early recognition of the importance of the role of the industrial chemist. In Britain the industrial chemist was not accorded his rightful status in the nineteenth century.

(c) *Capital commitments* were undertaken in the First World War because of the demand for armaments and munitions, and further investment was undertaken in the brief post-war boom to expand capacity. Consequently, the industry was saddled with debt and excess capacity when post-war demand declined. Rationalisation and modernisation of the industry were necessary in the 1930s.

(d) In the 1930s *changes in tariff policy* and *rearmament* helped the industry on the road to recovery so that 13 million tons of steel were produced in 1939. Expansion was brought to a halt in the Second World War and there was increasing reliance on the American industry, the output of which increased by more than 50 per cent.

33. Technology.

(a) *The Bessemer process.* In 1855–6 Henry Bessemer evolved a method of producing malleable iron and steel in a converter, but the process did not eliminate phosphorous from the iron. Bessemer established his own firm and successfully undersold his rival iron and steel producers. (William Kelly in the U.S.A. developed a similar process at about the same time.)

(b) *The Siemens–Martin process.* Sir William Siemens and Pierre Martin perfected a method of producing steel in an open-hearth furnace and this permitted a greater control over the steel-making process.

(c) *Gilchrist–Thomas basic steel.* This was a process developed for removing phosphorous from the iron. It could be applied to both the above methods of making steel, and made the use of phosphoric ores possible.

(d) The *significance* of the Bessemer, open-hearth and Gilchrist–Thomas processes was as follows:

    (i) Steel production was cheapened and an increase in the scale of construction of ships, bridges, docks, locomotives, etc., was made possible.

    (ii) To produce acid steel Britain had to import large quantities of non-phosphoric iron ore.

    (iii) The Gilchrist–Thomas process made it possible to use vast reserves of phosphoric ores in Europe and the U.S.A. and facilitated the rapid growth of the industry in the U.S.A. and Germany.

    (iv) In Britain in particular much capital equipment had to be scrapped, as steel and not iron became the end product. A rise of new technology put Britain at some relative disadvantage compared with the U.S.A. and Germany.

(e) *Chamber ovens and the cleaning of gases* produced in smelting to use as fuel in gas engines for powering rolling mills were important fuel-saving developments that Britiain was slow to utilise in the nineteenth century, and this put the industry at a cost disadvantage against rivals overseas. These techniques were adopted in the early years of the twentieth century.

(f) *Some increase in the size of blast furnaces* resulted in the First World War, and a large-scale physical reconstruction of the industry took place in the 1930s, when Stewart & Lloyds built the large Corby plant, the Lancashire Steel Corporation constructed the Orlam works, Guest Keen Baldwins rebuilt their Cardiff works and two continuous hot wide strip mills were constructed – one at Ebbw Vale by Richard Thomas & Co. and one at Shotton by John Summers (mills of this type had been in operation for a number of years in the U.S.A.). More than

£50 million was spent on these and other schemes, and steel-producing capacity increased by 2 million tons.

(g) *Research and free interchange of ideas* were fostered by the following bodies:

(i) The Iron and Steel Institute, founded in 1869.

(ii) The Iron and Steel Industrial Research Council, established in the 1920s.

(iii) The British Iron and Steel Research Association, formed in 1944, which by 1955 was spending £500,000 a year on research.

34. Markets, 1850–75.

(a) *Exports.* Britain was the world's main producer of iron and steel and exported more than a third of her total output of iron in the period; the peak year was 1872, when exports totalled 3·4 million tons. The U.S.A. was an important market for railway iron and Russia and India also absorbed substantial quantities of British iron for railway development. Germany imported pig iron from Britain.

(b) *Domestic markets.* Railway companies, cutlery firms, manufacturers of other metal goods and civil engineers provided the markets in Britain, and shipbuilding consumed an increasing amount of metal; in 1877 the tonnage of iron ships launched was approximately 400,000 and a few ships built of steel were launched also.

35. Rise of international competition, 1875–1913.

(a) *Exports:*

(i) Competition from the U.S.A., Germany and, to a lesser extent, Belgium grew so that even Britain's domestic market was invaded by German and American producers – especially the former. Cost advantages moved in favour of the U.S.A. and Germany, and by 1913 German exports were 33⅓ per cent greater in value than Britain's; the U.S.A. exported about two-thirds of the volume that Britain sold abroad. German production in 1913 was nearly double Britain's output and American production nearly doubled that of Germany.

(ii) The complementary aspects of industrialisation ensured a growth in absolute terms of British exports, but by 1913 competitive effects were causing conern among British iron and steel producers. World demand for railway materials remained high, but Britain's share of this trade was being nibbled at by the U.S.A. and Germany. Markets in these two countries were lost because of cost advantages achieved by the Americans and Germans, and because of tariff barriers; *e.g.* exports of tinplate from Wales to the U.S.A. were severely diminished by the tariff imposed in the latter country.

(b) *Domestic markets.* Demand for railway iron and steel continued at a high level both for new development and for replacement purposes. Nearly 6,000 miles (9,600 km) of new railroad were constructed in the period 1875–1913, although new development began to tail off after 1898. Demand for iron and steel for shipbuilding, manufacturing and construction increased. (The tonnage of ships built of iron and steel was 40,000 in 1899 and 950,000 in 1913.)

Producers of semi-finished steel found the domestic market being invaded by German producers by the turn of the century. In 1913 imports of iron and steel exceeded 2 million tons.

36. Market difficulties after 1918.

(a) *Marketing difficulties* were experienced both abroad and in the homeland in the inter-war period:

(i) Shipbuilding and heavy engineering experienced difficulties and cut back demand in the early 1920s, and in 1929 to 1933 especially. (In 1933 the tonnage of ships launched was less than 100,000.)

(ii) The industry was a relatively high-cost one. This, plus the tying of the pound to gold at a high exchange rate, made exports relatively expensive. Reorganisation in the 1930s went some way towards meeting the problem of costs.

(iii) In the 1920s lack of protective tariffs enabled some foreign rivals to dump products on the British market at prices that were below production costs. At the same time British producers had to endeavour to surmount foreign tariff barriers. (Cheap foreign crude steel, however, helped keep down costs of production in manufactures.)

(iv) World-wide depression in the years 1930–3 reduced the volume of world trade, but protective tariffs in 1932 helped preserve the home market and tariffs were reinforced by import quotas in 1935. Protection, however, cut off manufacturers from supplies of imported relatively cheap crude steel and exacerbated difficulties in export markets.

(v) Demand for railway iron and steel was for replacement purposes only—railway companies were in difficulties, too.

(b) *Compensatory factors were as follows:*

(i) The increased demand for steel from the motor vehicle industry.
(ii) The increasing cost of domestic electrical appliances.
(iii) Rearmament in the 1930s.
(iv) Bilateral trade agreements.

(c) *Home demand expanded enormously* after the Second World War and successful efforts were made to boost sales of exports, helped, to some degree, by devaluation of the pound in 1949.

37. Organisation of the industry.

(a) *Before 1914.* Economic and technological factors in the iron and steel industries operate in such a way that the optimum size of unit is extremely large. Integration of firms occurred in the nineteenth and early twentieth centuries: in 1914 there were approximately a hundred producers of pig iron and one hundred producers of steel. Amalgamations were centred around steel making, shipbuilding and armament production. Average unit size was small compared with some American firms.

(b) *1918–39.* Amalgamation speeded up after the First World War, much of it a result of pressure by the Bank of England. By 1926 the number of pig-iron producers was seventy and the number of steel producers seventy-five; in 1930 70 per cent of the iron and steel output was produced by twenty firms; and by 1932 ten groups possessed about half the pig-iron productive capacity and 60 per cent of steel capacity. (Even so, the United States Steel Corporation had a productive capacity more than treble the total capacity of the ten groups. The capacity of the Bethlehem Steel Corporation, too, exceeded that of the groups.) Amalgamations and reorganisation were fostered by the Import Duties Advisory Committee, set up in 1932, and the British Iron and Steel Federation, which was able to persuade the industry to participate in fixing prices.

(c) *After 1939.* During the Second World War control was imposed through the Ministry of Supply, and, in 1945, the Labour government announced plans for nationalisation of the industry with a concentration of production in fewer, larger units, and in 1949 the Iron and Steel Corporation was formed, but it was very short-lived. Output per furnace (blast and open-hearth) increased by about 50 per cent between 1946 and 1954 and labour productivity rose.

GROWTH OF NEW INDUSTRIES

38. Impact of technology.

(a) *Nineteenth-century innovations* which affected life in industrialising countries most of all were those in:

(*i*) steam-engine technology as applied on railroads, in shipping and in industry generally;

(*ii*) metallurgy, enabling large quantities of iron and then steel to be produced at economic prices;

(*iii*) the machine-tool industry, facilitating the accurate production of parts for machines;

(*iv*) the textile industries, which mass-produced vast quantities of cloth and yarn at relatively low prices;

(*v*) industrial chemistry.

(*b*) *Other factors which changed life* in the twentieth century rather than the nineteenth were:

(*i*) the development of the internal combustion engine;

(*ii*) the discovery of methods of producing electricity on a fairly large scale;

(*iii*) developments in inorganic and organic chemistry (the latter provided a basis for a development of new industries after 1918);

(*iv*) the development of communication media such as radio, television and the telephone service;

(*v*) the production of domestic electric appliances.

(*c*) *New industries arose*, based on discoveries made in the second half of the nineteenth century and the first half of the twentieth century – industries that had as much impact on life in Britain as the old staple industries had between 1850 and 1930.

THE MOTOR INDUSTRY

39. Conditions favourable for growth. In the nineteenth century conditions facilitating the growth of the motor vehicle industry developed:

(*a*) Alloy steels combining lightness and strength were evolved.

(*b*) Mineral lubricants and petroleum had a high production potential.

(*c*) Electrical engineering advanced.

(*d*) In the Midlands a diversity of light engineering industries arose – industries that could be adapted for producing parts for motor vehicles.

(*e*) Rubber technology improved.

(*f*) *Per capita* national income increased.

40. Growth and organisation.

(*a*) *Before 1914*:

(*i*) Motor cars were usually made to a customer's order and firms were small. In 1914 there were about 200 producers with an output totalling approximately 30,000 vehicles.

(*ii*) The scale of production was much smaller than in the U.S.A., where by 1912 seven firms were producing more than 500,000 vehicles a year. American producers aimed at mass markets and mass-produced standardised relatively inexpensive vehicles; British producers had a more limited wealthy market in view.

(b) *Change in emphasis after 1918:*

(*i*) The inter-war period was one of expansion in output and rationalisation. The number of firms fell to ninety-six by 1922 and to twenty by 1939, with six firms dominating production. Output increased to more than 300,000 vehicles per annum by 1927 and exceeded 500,000 in 1937. There was a decline in the Second World War but in 1950 production exceeded 750,000 units and in 1963 was more than 2 million units.

(*ii*) After the Second World War concentration of production into fewer firms continued. In 1969 it was largely in the hands of four large groups: British Leyland, Ford, General Motors and Rootes.

(*iii*) There was a concentration of location in the Midlands and the London area.

(*iv*) Employment was provided for an increasing number of workers. Between 1923 and 1939 the labour force doubled, being in excess of 300,000 in the later year.

(c) *1918–39.* Weaknesses existed in the industry in the inter-war period:

(*i*) Firms were of a relatively small size.

(*ii*) Firms produced a diversity of models and a diversity of engines.

(*iii*) The taxation system channelled home demand and production into cars with small, high-revving engines, which were not favoured in export markets with large continental areas.

Because of small firms, and because of the large numbers of models and engines, costs were higher than necessary. Economies of scale could not be achieved to the degree necessary, and production runs were insufficiently long.

41. The effect of the great depression, 1929–32. American motor vehicle output fell from 5·4 million vehicles to 1·4 million between 1929 and 1932. Britain's output was only 239,000 in 1929 but it remained at about this level until 1933, when it began to increase. The reasons for the great depression having little adverse effect on production and for the increase in production from 1933 were as follows:

(a) Real *per capita* income continued to increase in Britain, despite high unemployment in the staple industry areas.

(b) Exports of vehicles did not form a great proportion of total output – about 10 per cent – and the industry was not greatly susceptible to fluctuations in demand abroad. (The motor-cycle industry was badly affected: in 1929 45 per cent of output was exported and production was severely curtailed by the depression.)

(c) Fuel taxation policies of governments overseas increased the attraction of the small economical British cars. Between 1929 and 1937 exports more than doubled, but remained at about 10 per cent of total output.

(d) The industry was catering for "new" demand and not "replacement" demand.

ELECTRICAL ENGINEERING

42. Reasons for growth. In 1880 electrical engineering was virtually non-existent. In 1940 much of Britain was provided with electricity from a national-grid system, which after 1945 was extended to cover the whole of England, Scotland and Wales. The production and use of electricity became important in the twentieth century for the following reasons:

(a) It provided industry with a source of power that freed it from location near coalfields: the Midlands and the south grew in industrial importance.

(b) It enabled the telegraph, the telephone, radio and television to be developed as means of communication: a new era of mass communication was ushered in.

(c) The production and servicing of domestic electrical appliances provided employment for thousands of workers.

43. The period of greatest growth. The main growth was in the years 1925 to 1939:

(a) In 1925 the industry was relatively backward, with a variety of voltage systems and a large number of suppliers of current (about 440). This made large-scale production of electrical goods and apparatus difficult.

(b) In 1926 several important developments took place:

(i) The *Electricity Supply Act* was passed.
(ii) The Central Electricity Board was set up.

(*iii*) The national-grid system was formed, fed by a small number of large generating stations. Results were: (1) an increase in generation (*e.g.* 1925, 6·6m. kW; 1939, 26m. kW; 1950, 56m. kW; 1964, 164m. kW); (2) the standardisation of supply; (3) a great expansion of electrical engineering – 367,000 were employed by 1937; (4) the electrification of some railroads; (5) an increase in the number of users of electricity (*e.g.* 1920, 730,000; 1938, 9 million).

(*c*) Generation of electricity was not affected by the depression of 1929–33 – demand continued to grow rapidly in the 1930s.

THE ENTREPRENEUR AND THE ECONOMY, 1875–1914

44. Controversy. The enterprise of entrepreneurs and managers in industry and commerce after 1875 has been subject to considerable criticism by both contemporaries and recent writers. Concern was expressed because of the relative decline of British industry after 1875: the U.S.A. and Germany began to overtake Britain in industrial output. This was due to some natural advantages, such as the size of the domestic market and the location of supplies of raw material but, additionally, some observers consider it was due to complacency and lack of enterprise on the part of British entrepreneurs, who were slow to adopt new machines and new processes in production, and who did not analyse critically enough their new methods of management and organisation.

(*a*) *Pessimistic views* have been expressed about entrepreneurs in the staple industries and the chemical industry. In the 1880s Andrew Carnegie criticised British manufacturers of iron and steel products for using out-of-date plant, particularly in converting pig iron to steel (*see* **45** below).

(*b*) *More optimistic views* about entrepreneurs are expressed in relation to the burgeoning consumer-goods industries and the distributive trades (*see* **46** below).

45. The pessimistic view.

(*a*) British entrepreneurs in the years 1875–1914 are subject to criticism in that they take some of the blame for the deceleration in economic growth:

(*i*) They had contempt for change and innovation.

(*ii*) They worked according to tradition and only a small part was accorded to science and research in production.

(b) Other criticisms levelled at entrepreneurs include the following:

(i) Iron and steel producers failed to keep abreast of technical changes in the industry and operated at relatively high costs.

(ii) Coal producers are considered to have neglected investment in labour-saving machines, and are accused of passing on high costs to customers.

(iii) Entrepreneurs in the tinplate, cotton and shipbuilding industries are accused, too, of failure to keep abreast of modern innovation and of operating at costs that were too high.

(iv) In particular, industrialists are attacked for failure to innovate in the newer industries – the growth points of the post-1918 period. It is observed that entrepreneurs were insufficiently orientated towards science and technology, and were markedly reluctant to employ university graduates. This contrasts strongly with entrepreneurial attitudes in the U.S.A. and Germany at that time.

(v) Firms remained too small and were dominated to a great extent by family control. There was a reluctance to appoint progressive managers and to seek capital from wider sources.

46. The optimistic view.

(a) The more optimistic school argues that the fortunes of the heavy industries do not necessarily reflect the development of the economy as a whole. The growth points were in the consumer-goods sector where great changes were effected between 1851 and 1911. Beginnings of great industrial empires were founded by entrepreneurs such as Beecham, Boots, Cadbury, Colman, Courtauld, Lever, Lewis, etc. There was no lack of enterprise by entrepreneurs such as these, who formed part of a group of entrepreneurs who developed:

(i) new techniques of selling;
(ii) new methods of organisation in wholesaling and retailing (e.g. Sainsburys);
(iii) new trades;
(iv) new types of consumer goods;
(v) large-scale production requiring national markets and the creation of markets through advertising, etc.

(b) New firms in light engineering are shown to have displayed considerable initiative in exporting and adopting new machines and methods of organisation.

(c) Newspaper entrepreneurs were not lacking in initiative and built up large business empires.

47. Summary of views.

(a) Whilst entrepreneurs in the staple industries might have been complacent and unenterprising a relative decline was almost inevitable because:

(i) Britain had no enduring comparative advantage in such industries;

(ii) the domestic market was small compared with those of Germany and the U.S.A.;

(iii) indigenous raw materials were more inadequate in amount (except for coal) than in the U.S.A.;

(iv) the early start was a disadvantage: there was heavy investment in capital equipment and a reluctance to scrap when plant became obsolete, and a reluctance to rationalise production to become more competitive.

(b) There was a marked lack of urgency in building a broadly based system of state education, with a consequent lack of provision for the study of the natural sciences. The number of graduates in science, technology and mathematics was woefully small – even in 1913. Britain lagged behind the U.S.A. and Germany in the scientific–technological growth industries (e.g. chemical, techno-electrics, machine-tool and automobile industries).

(c) Whilst succeeding generations of entrepreneurs might have been less enterprising than their forebears there is plenty of evidence to illustrate enterprise and initiative, especially in the consumer-goods industries. Businessmen tend not to lack initiative.

PROGRESS TEST 8

1. How important was the cotton industry in the industrial revolution? (1, 2)

2. What was the economic significance of changes in British company law? (4–7)

3. Examine business consolidation in England and Wales before 1914. (8–12)

4. Outline British industrial concentration after 1918. (13–16)

5. Why was the British coal industry in economic difficulties after 1918? (19, 20)

6. Discuss the development of the cotton industry between 1850 and 1950. (27–31)

7. What factors aided the growth of steel output in Germany and the U.S.A. after 1870? (32)

8. What marketing difficulties did British iron and steel producers encounter after 1900? (35, 36)

9. Discuss the organisation of the British iron and steel industry in the twentieth century. (**37**)

10. Consider the growth of the motor vehicle industry in the inter-war period. (**40, 41**)

11. To what extent were entrepreneurs in Britain too complacent in the years 1875 to 1914? (**44–47**)

CHAPTER IX

INDUSTRY IN THE U.S.A.

THE GROWTH OF INDUSTRY

1. The leap forward, 1850–1900.

(*a*) In 1850 the value of output of manufactures in the U.S.A. was less than that of each of the countries Britain, Germany and France. By 1900 American industry was producing as much as the industries of all three countries combined.

(*b*) At the outbreak of the Civil War the nation was still primarily an agricultural one: by 1890, despite a vast increase in agricultural output, industry contributed more to the national income than did agriculture. Between 1860 and 1890 the number of workers in manufacturing and construction increased fourfold.

(*c*) The U.S.A., by 1900:

   (*i*) had the highest *per capita* income in the world;
   (*ii*) was the largest producer of agricultural products;
   (*iii*) produced about a third of the world's manufactures.

2. Reasons for the leap forward.

(*a*) The domestic market became of great size and considerable depth. The population increased by 1 million per year from 1860 to 1910 and income per head increased rapidly.

(*b*) Industry benefited from protective tariffs.

(*c*) Supplies of capital increased. Foreign capital was of some significance, but became relatively less significant as the century passed by: indigenous capital was of far greater importance.

(*d*) The labour force grew, although wages as a whole remained relatively high. Immigration was an important factor.

(*e*) Transportation developments improved factor mobility and were of great cost-reducing effect.

(*f*) Technological progress was rapid. Labour-saving machines, chemical technology and the development and use of new sources of energy (oil and electricity) transformed the industrial scene.

(g) Scientific methods of management and organisation were evolved, aided by innovations in communications and office equipment and permitting an increase in the size of business units which resulted in the achieving of economies of scale.

(h) Entrepreneurs showed great enterprise.

3. Growth, 1900–50.

(a) *Industry expanded enormously* over the half-century. Comparing 1900 with 1950:

  (i) the number employed in manufacturing increased nearly threefold;

  (ii) real product per man-hour more than trebled;

  (iii) the gross national product increased almost fivefold;

  (iv) the average weekly earnings of production workers in manufacturing increased by more than 500 per cent.

(b) *Setbacks* to industrial growth occurred in 1907, 1921–2, 1930–3 and 1938.

(c) The *pace-setters* in the twentieth century have been the automobile industry, techno-electrics, aircraft and space research industries, and the chemical industries. Since 1940 military demand has been an important factor in stimulating economic growth.

THE ORGANISATION OF PRODUCTION: GROWTH IN THE SIZE OF FIRMS

4. Development of corporations. As in Britain some suspicion of incorporation with limited liability existed, but in the 1840s state legislatures began to enact laws permitting incorporation with limited liability by mere compliance with conditions laid down in general terms. By 1875 special charters of incorporation were defunct in most states.

By 1920 corporations employed almost 90 per cent of the labour force engaged in manufacturing and produced nearly 90 per cent of manufactures, but corporations comprised only about 40 per cent of the total number of manufacturing firms.

5. The combination movement before 1914. After the Civil War combination of firms occurred and is often referred to as "business consolidation." The movement was intense between 1897 and 1905, but concentration of production had reached a high level in some industries by 1890.

Firms combined in several ways:

(a) Mergers and take-overs were effected (e.g. in iron and steel).

(b) Pooling arrangements were made and markets divided on a basis of output or territory or profit sharing (e.g. in salt, coal, steel rails, whisky, explosives, etc.).

(c) Trusts were formed: firms were linked together by trustee devices to overcome weaknesses of pooling arrangements (e.g. Standard Oil Company, American Sugar Refining Company, etc.)

(d) Holding companies were formed: New Jersey State became a centre for this (e.g. Standard Oil Company (formed in 1899), International Mercantile Marine Company, etc.).

6. Reasons for the growth in size in the nineteenth century.

(a) *Rapid expansion of concentrated market areas,* because of transport innovations and population growth, occurred. Mass markets developed.

(b) *Technological change was rapid.* One innovation led on to another and facilitated the rise of large business units.

(c) *Limited liability* and the corporate form of business unit found relatively ready acceptance in the U.S.A.

(d) *Fear of cut-throat competition* induced firms to combine into large organisations, especially in industries where overhead costs were high, e.g. iron and steel, railroads, sugar production.

(e) *Federal patent laws* enabled firms to collect numbers of patents and so acquire control of large sectors of manufacturing.

(f) *Investment bankers* such as J. P. Morgan promoted business consolidation partly to eliminate cut-throat competition and partly to reap monopolistic profits from mergers.

(g) *Vertical integration* held the promise of low costs and ensured supplies of raw materials and markets for products.

7. The effects of consolidation before 1914.

(a) *Effects on prices.* Controversy about the effects of large units and monopolistic trends on prices exists. The Bureau of Corporations of 1907 and the National Industrial Conference Board of 1929 produced conflicting findings. The former concluded that the Standard Oil Company used monopoly power to force up oil prices; the latter stated that prices of products of firms in large combinations rose less than prices of products from firms not affected by combination. One view is that in the short run prices were forced up but in the long run economies of scale kept prices lower than they would have been if units had remained small.

Prices probably stabilised in the long run and competition turned from price to service.

(b) *Concentration of power*. Power became concentrated in the hands of directors and managers.

(c) *Anti-trust movement*. (*See also* **9** below.) An anti-trust movement arose. State and federal authorities intervened to control the large industrial groups:

> (i) 1890. The *Sherman Anti-Trust Act* was passed.
> (ii) 1902–7. Roosevelt followed a "trust busting" policy.
> (iii) 1914. The Federal Trade Commission was formed to investigate corporations.
> (iv) 1914. The *Clayton Act* was passed to prohibit price discrimination and control the build-up of interlocking directorates.

Before 1914 anti-trust legislation was not very effective because of inadequate administrative provisions, inadequate penalties and inconsistency on the part of the Supreme Court.

8. Consolidation of industry after 1918.

(a) *Large combinations* were evolved after 1918. The movement was facilitated by the following factors:

(i) The attitude of the Supreme Court towards large industrial groups. Concentration of control seemed acceptable, but actions against restraint of trade were pursued.

(ii) Changes in technology, so that one unit of management could control a larger industrial organisation.

(iii) The need to have large amounts of capital for industry as plant became more complex.

(iv) The need to minimise competition because of large capital requirements in industry. International cartels arose in sulphur, dyestuffs, tungsten carbide, etc.

(v) Fairly widespread industrial prosperity in the 1920s. Business confidence was high.

(b) A pyramid structure of *holding companies* was utilised in the 1920s to bring about large-scale consolidation of firms. The structure extended into the automobile industry, food packing, banking, distribution, etc., as well as being common in railroads, entertainment and utilities. Such a business structure proved very fragile in the 1929 Wall Street collapse. In the chemical, automobile and oil industries a multi-divisional group structure was evolved, also, and this proved more durable than the pyramid holding company structure.

(c) *The impact of the New Deal* on business consolidation is not easy to measure:

(i) The *National Industry Recovery Act* encouraged the growth of oligopoly and near-monopoly. Codes of "fair competition" were enforced by the government, and many codes were essentially cartel-type agreements to maintain prices at a desired level.

(ii) In 1937 Thomas Arnold revived the anti-trust crusade under the *Sherman Act* to increase competition and to reduce prices.

(d) In 1950 the situation was as follows:

(i) By 1950 some 600 companies transacted the bulk of the nation's business. Each large group had assets exceeding $50 million. The residue of the market was shared by about 4 million smaller enterprises.

(ii) The professional entrepreneur became dominant in American business. Considerable attention was paid to the education of business managers in the 1950s.

9. The anti-trust movement.

(a) The *Sherman Anti-Trust Act*, 1890, was passed by Congress:

(i) to outlaw restraint of interstate commerce whether it be by contract, combination or trust;

(ii) to curb attempts to establish monopolistic positions in commerce.

Little use was made of the Act from 1890 to 1905, when Theodore Roosevelt took office, committed to a policy of enforcement of the anti-trust law. But in the next twelve years, under the Roosevelt, Taft and Wilson administrations, 133 cases were brought. However, the trend in industry towards the monopolistic or oligopolistic firm was not halted by the *Sherman Act*. Only two large-scale mergers were subject to adverse decisions – those resulting in the Standard Oil Co. and the Consolidated Tobacco Co.

(b) The *Clayton Anti-Trust Act*, 1914, supplemented the *Sherman Act* and a Federal Trade Commission was set up to enforce the Act through the courts, but the *Sherman Act* remained the base of the anti-trust policies. The Act was of only limited effectiveness in the 1920s because of the attitude of the Supreme Court.

(c) The *National Industry Recovery Act*, 1933, effectively suspended the *Sherman Act* for a brief time, but in 1937 Arnold revived enforcement of the *Sherman Act* and more effective steps to curb the monopolistic tendencies of large combinations were taken between 1940 and 1948 than ever before.

CAPITAL FOR INDUSTRY

10. Methods of raising capital.

(*a*) Capital was provided by the proprietors or partners in a firm.

(*b*) Profits were accumulated and then ploughed back into the industry.

(*c*) Loans were raised from various sources, such as banks.

(*d*) "Blind" capital was utilised by the sale of company shares and bonds through the medium of the stock exchanges.

11. The sale of industrial securities.

(*a*) Prior to the Civil War the main supplies of shares and bonds came from the railroad and canal companies and from state and national governments. The *chief buyers* were banks and wealthy individuals, and the foreign market was important. After 1864 an increasing supply of railroad and other industrial securities appeared on the market and induced a growing specialisation of function among operators in the capital markets of the U.S.A.

(*b*) *Underwriters of security issues* became prominent in the marketing of securities. In the Civil War Jay Cooke & Co. took the whole of the issues of certain government securities and disposed of them on the market by means of advertising campaigns. After 1870 groups of financiers underwrote issues of industrial securities.

12. Role of investment banks.

(*a*) Banking firms such as those of J. P. Morgan & Co., Kuhn, Loeb & Co., and the National City Bank became active in the field of *underwriting*, and launched out into the floating of industrial securities in the 1860s and 1870s.

(*b*) Investment bankers soon began to exert influence and control over *corporations* whose issues they underwrote because:

   (*i*) funds raised were not used wisely to promote growth of the company;

   (*ii*) reorganisation of companies advised by bankers was not always effected;

(*iii*) the reputation of investment banks suffered when firms whose issues they had floated did not prosper according to expectations.

(*c*) Control over institutions purchasing securities was sought, too. Thus there was a welding of investment banks, insurance companies and commercial banks into integrated structures.

(*d*) *Holding company* and *trust devices* were used frequently by investment bankers to acquire control of firms and many amalgamations were thus effected. For example, J. P. Morgan & Co. were actively concerned with the formation of the United States Steel Corporation (capitalised at more than $1·2 billion), the International Harvester Co., the International Mercantile Marine Co. and the General Electric Co.

(*e*) Business on the *New York Stock Exchange* boomed in the late 1890s because of the consolidation movement instigated by investment bankers. Dealings in shares multiplied fourfold between 1896 and 1901.

(*f*) *Concentration of underwriting* of issues developed among investment banks:

  (*i*) within New York City; and
  (*ii*) among a relatively few firms.

Between 1936 and 1945, approximately twenty investment banks handled nearly 70 per cent of all the business contracted in relation to new issues and about 240 did the rest – excluding the raising of capital for small businesses.

13. Investment trusts. In the 1920s investment trusts were devised to harness the capital of relatively small investors. Some trusts were created by investment bankers and about 400 were in existence by the end of the 1920s.

14. Savings banks and insurance companies. Funds for small savers were (and still are) channelled into investment by savings banks and insurance companies. In 1950 more than 33 per cent of insurance companies' reserves were invested in bonds and stocks and these institutions have become of great significance in the stock market, often by-passing investment bankers.

15. Importance of foreign capital.

(*a*) *Foreign investors* were important suppliers of capital before 1860, especially in facilitating canal and railroad developments, through the purchase of state bonds. Estimates of foreign capital invested in the U.S.A. vary, but it can be concluded that:

(*i*) the U.S.A. was an international debtor on capital account until 1914–18;

(*ii*) from 1918 to 1929 the U.S.A. was a substantial creditor nation;

(*iii*) between 1929 and 1939 the gap between capital exports and imports diminished to about $1 billion.

(*iv*) from 1941 there was a substantial flow of capital from the U.S.A.: government overseas aid alone amounted to nearly $40 billion between 1945 and 1952.

(*b*) Capital imports *from 1780 to 1861* are estimated to have been between $400 million and $500 million. Investments in the U.S.A. were made by German, Swiss, Dutch and British investors. The British money market and Anglo-American merchant bankers were important media through which issues of American securities were sold, but by 1860 the American money market was growing in importance in supplying capital for industrial development.

(*c*) *From 1860 to 1914* capital imports grew to more than $4 billion. Foreign capital, especially from Britain, was important in aiding railway development before 1890, but most British railway investors were bought out in the consolidation movements of the 1890s.

(*d*) Capital exports increased during the period 1860–1914 as domestic capital supplies grew, so that by 1914 some $3·5 billion were invested abroad.

THE IRON AND STEEL INDUSTRY

16. Production of iron and steel. Technological developments (as in Britain) influenced the nature of production, the main end-product changing from iron to steel. Whilst Bessemer steel bulked large in total British output, open-hearth steel was more important in the U.S.A. from about 1908 onwards. In 1940 more than 90 per cent of total output was open-hearth steel.

A dramatic increase in output occurred from 1870:

(*a*) Pig-iron output grew from 2·5 million tons in 1873 to more than 40 million tons by the end of the First World War.

(*b*) Steel production by 1910 exceeded 25 million tons and topped 55 million tons in 1929. In the 1929–32 depression output fell catastrophically to 14 million tons in 1932, but then recovered strongly to more than 50 million tons by the end of the 1930s, when American production exceeded the total output of Japan and the major western European producers. Reasons for the recovery after 1932 were as follows:

(*i*) New Deal policies – deficit financing, the Tennessee Valley project, the *Reciprocal Trade Agreements Act*, etc.

(*ii*) Rearmament in the 1930s.

(*iii*) An increasing demand for U.S.A. steel from Britain and her allies during the Second World War: in 1944 output exceeded 90 million tons.

17. Iron ore. Vast reserves of iron ore were discovered in the nineteenth century:

(*a*) Until 1850 the upper reaches of the Monongahela River were one of the main ore-mining areas. The ore was processed locally and then fed the furnaces in Pittsburgh.

(*b*) From the 1850s the Lake Superior ores were mined, but exploitation of the ores increased after the Gilchrist–Thomas process was developed. The area became the chief ore-producing region: in 1946 more than 80 per cent of iron ore produced in the U.S.A. came from here

(*c*) Other ore-producing regions of note in 1950 were the Birmingham region and the Adirondacks.

(*d*) By 1950 American steel companies had acquired interests in ore deposits abroad – in Canada, South America and Africa.

18. Bessemer and open-hearth processes.

(*a*) The *Bessemer* method of steel production was introduced in the 1880s after the resolving of the Kelly–Bessemer conflict over patent rights, and in 1867 steel rails were produced. By 1900 steel-rail production was approximately 3 million tons.

(*b*) *Open-hearth* steel was a higher-grade product than Bessemer steel and gradually ousted the latter. The Gilchrist–Thomas process made this evolution possible.

19. Electric furnaces. During the First World War the use of electric furnaces was boosted because of demand for special alloy steels. Electric furnaces provided greater control over temperature and chemical reactions, and were built in areas where hydro-electricity was available. Electric furnaces displaced crucible and cupola methods of steel production.

20. Alloy steels. By 1940 11 million tons of alloy steel were produced per year. These are steels with special additives to provide greater strength, better resistance to corrosion and resistance to higher temperatures. Growing sophistication of machines made such metallurgical improvements necessary.

21. Rolling mills. Rails, plates, sheets, etc., were demanded in such quantities that the installation of rolling mills increased rapidly. By 1910 16 million tons of rolled steel were being produced per year: in 1929 41 million tons were produced. Rolling mills produced steel rails, plates, sheets, girders, etc. Continuous strip mills were introduced in the 1920s and had a great cost-reducing effect. Cheap sheet steel made possible the mass production of consumer goods such as motor cars, domestic electrical appliances, toys, etc.

22. Size of blast furnaces. The U.S.A. had easily surpassed Britain in building big blast furnaces by 1900. Average output per furnace increased more than twentyfold between 1860 and 1900 and resulted in considerable cost advantages for American producers as compared with their British counterparts.

23. Growth in size of units. Technological change was accompanied by organisational change because of:

  (a) the increasing amounts of fixed capital required;
  (b) the need to keep costs low by reaping the benefit of economies of scale;
  (c) cut-throat competition;
  (d) the activities of investment bankers seeking profit from such mergers.

Vertical and *horizontal integration* occurred between 1870 and 1907. A great number of mergers took place in the 1890s. Between 1898 and 1901 twenty-one large combinations were formed.

24. The United States Steel Corporation. The amalgamation movement before the First World War culminated in the forming of the U.S.S.C. in 1901, in which eleven large groups were absorbed in an organisation capitalised at over $1·4 billion, and which took over five more companies in the next six years, giving the group control of the bulk of production of steel tubes, barbed wire, fencing wire, etc.

25. Mergers after the First World War. Two large groups were formed by mergers in 1922 and 1923:

(*a*) The Bethlehem Steel Company group.

(*b*) The Youngstown Sheet and Tube Company group.

These two groups, plus the giant U.S.S.C., controlled nearly 70 per cent of the U.S.A.'s steel-ingot capacity, the rest being controlled by four large and about fifty independent companies.

26. Domestic market. More than 90 per cent of total production before 1939 was consumed in the domestic market. The American industry, therefore, contrasts strongly with the British industry in this respect. Main consumers of steel were:

(*a*) railroads;

(*b*) mineral industries;

(*c*) motor vehicle works;

(*d*) farmers;

(*e*) food-container manufacturers;

(*f*) machine producers;

(*g*) a host of miscellaneous producers.

Main areas of consumption before the Second World War were Pennsylvania, Illinois, Ohio and Michigan, which together consumed more than 60 per cent of total output, but consumption of steel was, of course, widespread over the whole of the U.S.A.

27. Export markets. Markets of chief importance were Canada and Japan, followed by South America and Central America, but European nations provided important markets during the First World War, the 1920s and the 1940s. In 1900 iron and steel manufactures constituted about 15 per cent in value of the total of American exports; Britain and the Empire were markets of great significance. In the 1920s Europe purchased about a third of total American exports of manufactures, especially iron and steel manufactures.

During the Second World War a vast quantity of American materials was conveyed to Britain and the Allied nations, and after 1945 there was still considerable reliance on American industry until war-shattered Europe could recover.

THE COAL INDUSTRY

28. Output.

(*a*) Demand for coal grew rapidly after 1870, when a change-over to coal as a fuel became more general on the railroad and in the iron industries. In 1850 the total output was only about one-eighth of British output: in 1918 it was more than treble the output of British mines, being about 700 million tons.

(b) Output declined in the inter-war period. Like the industry in Britain, the American coal industry had to endure competition from other fuels – oil and natural gas – and great economies in the use of coal as a fuel were made.

(c) The 1929–32 depression affected the coal industry severely: unemployment was high.

(d) Demand for coal increased in the Second World War: in 1944 output approached the 1918 total. After 1945 demand again receded in face of competition from oil, natural gas and hydro-electricity.

29. Organisation of the industry. The anthracite and bituminous coalfields can be considered separately:

(a) Many *anthracite coal-mines* were gradually taken over by railroad companies between 1870 and 1910. By 1920 producers numbered less than 200 and competition was minimised by the arrangement of "pools" with the effective operation of a quota system. A high proportion of coal-mining was mechanised by 1920: further mechanisation increased productivity between 1920 and 1950. Demand for anthracite declined as railroads met economic difficulties and as oil replaced coal in mercantile fleets.

(b) *Bituminous coal-mines* present a different picture. Organisation was more complex. Some coal-mines were parts of large industrial complexes with assured markets for coal, but many were independent units: there were more than 12,000 producers at the end of the First World War. Rationalisation in face of falling demand in the inter-war period was difficult to effect. Unemployment was high: many mines had to close down because of the fall in coal prices. The New Deal policies did little to help this section of industry out of its difficulties because of the large number of producers involved. Industrial relations were not good: the Mineworkers' Union pressed for nationalisation of the coal industry. Strikes in the coal industry and other basic industries led to the passing of the *National Labour Relations Act* in 1947.

30. Markets. Only a small proportion of output was exported: the domestic market was the prime market. In 1900 90 per cent of domestic fuel requirements were met by coal, but by the end of the First World War the proportion was less than half because of the rise of competing fuels. Large reserves of oil and natural gas were exploited so that although coal-mining costs were kept down by increasing mechanisation the domestic market shrank after 1920 except for a brief revival in the Second World War.

THE AUTOMOBILE INDUSTRY

31. Growth, 1900–20.

(a) *Output increased rapidly.* In 1900 it was about 4,000 units per year but in 1920 it exceeded 2 million. Its rapid rise was due to:

(i) the large domestic market with a high *per capita* income;

(ii) the need for a convenient form of transport in rural areas remote from railroads;

(iii) technological advances in the metallurgical, oil, rubber and glass industries;

(iv) the existence of a sophisticated machine-tool industry;

(v) the availability of adequate supplies of capital;

(vi) a willingness on the part of the population to accept standardised mass-produced products.

(b) *Mass-production techniques* were adopted by the industry and notably by Henry Ford, who produced more than 15 million Model T Fords before 1925 and at the same time reduced prices by more than 65 per cent. Minute divisions of labour, conveyor-belt flow systems, sophisticated stock control and up-to-date management techniques revolutionised production: costs fell dramatically and productivity leapt forward. Vertical integration was utilised to increase economic efficiency.

32. The prosperous 1920s.
The U.S.A. experienced a remarkable economic boom between 1922 and 1929 and the automobile industry was a main factor in promoting economic growth. Until 1925 demand for vehicles was chiefly "new" demand, but after 1925 replacement demand increased and compelled producers to initiate design changes.

Economic effects of the expansion of the industry were as follows:

(a) Road-building programmes were undertaken: nearly $2 billion per year was invested in roads by the end of the 1920s.

(b) Construction of urban suburbs took place.

(c) Demand for steel, oil, glass, rubber, etc., was stimulated.

(d) A large-scale construction of filling stations and service stations occurred.

(e) Nearly 4 million jobs were created and mobility of labour was improved.

(f) Capital investment increased.

(g) Distribution of goods speeded up, releasing capital from the distribution network.

(h) Research in the oil, petroleum, rubber, glass and metallurgical industries was intensified.

Annual output in 1929 exceeded 5 million vehicles and more than 25 million vehicles were registered by the end of 1929. The industry was the most important industry in the U.S.A. by the end of the 1920s, when the U.S.A. produced more than 80 per cent of the world's output of motor vehicles.

33. The years 1929–50. In the depression of 1929–32 output fell to 1·4 million units. Recovery was fairly rapid to 1937, when a recession set in, but this trend was reversed in 1939. In the Second World War lend-lease policies, and after 1945 Marshall Aid, plus rising real incomes in the U.S.A., stimulated demand for motor cars and trucks. Demand for automobiles outreached supply, even in 1948, when output exceeded 5 million vehicles.

34. Labour relations. During the 1930s the C.I.O. organised labour in the industry so that by 1941 all the motor corporations recognised unions for collective bargaining purposes (*see* XIV, **36** (c)).

35. Organisation. Scale of production and standardisation of parts are important facets of the industry. Advanced systems of production can be used economically only where output is to be large. Concentration of production in the hands of a few large organisations helped minimise costs: by 1939 90 per cent of output was produced by three companies and 50 per cent of sales consisted of sales of three models only, giving production runs of more than 350,000 each. This gave the American industry important cost advantages over the British motor industry except in the 1929–32 depression, when output fell well below optimum levels.

ELECTRIC MACHINERY AND ELECTRIC APPLIANCES

36. Electric power. In 1919 30 per cent of powered machinery in industry was electric: by 1925 more than 70 per cent of total factory power was supplied by electricity. Almost all this growth took place after 1900 because of the following factors:

(a) Technological advances enabled electricity to be generated in greater quantities at lower costs.

(b) Reliable electric motors were produced.

(c) Problems of transporting electricity were solved.

(d) There was a demand for a source of power with convenience and cost advantages for users and which made industry "footloose" in respect to coalfields.

Between 1919 and 1929 output of electric power and light more than trebled. In 1939 production had reached 161 million kW and exceeded 3 billion kW by 1950.

37. Effects of the rise of the industry. The development of the electrical industry had important economic effects after 1919:

(a) Machine-tool operations became more flexible. The use of hand tools was boosted.

(b) Demand for household electrical appliances mounted rapidly: electric irons, washing machines, cleaners and refrigerators revolutionised the lives of housewives in the U.S.A. There was, therefore, a rise of new manufacturing industries to satisfy demand titillated by advertising campaigns. By 1930 the electrical industry was producing goods at a rate of over $2 billion per year.

(c) Hire-purchase facilities expanded to enable demand to increase, but this induced more instability into the economy because such demand was not constant.

(d) Manufacturers became important purchasers of electricity from public utility companies, which tended to merge into large-scale units. In 1900 industry provided most of its own power: by 1939 60 per cent of power was purchased.

(e) Trends towards mass production were intensified by the supply of sophisticated automatic instruments to control operations.

(f) Hydro-electricity provided convenient, relatively cheap power for use in aluminium, magnesium, steel, chemical and paper-making industries.

38. The radio industry.

(a) Because of *technological advances* in electrical engineering and the existence of a *large consumer market,* in both depth and width, the radio industry surged into faster growth. The value of output grew fortyfold, to about $400 million, in a decade.

(*b*) The industry facilitated the growth of more intensive mass advertising campaigns.

(*c*) *Hire purchase* was an important factor in enabling demand to grow rapidly.

39. Organisation. By 1929 the electrical industry was controlled largely by eleven groups.

INDUSTRY IN BOOM AND DEPRESSION, 1922–32

40. Economic boom, 1922–9. A tremendous upsurge in the economy occurred between 1922 and 1929 (with minor setbacks in 1924 and 1927). Industrial output nearly doubled in value, *per capita* real income rose by 40 per cent and the national income increased by 50 per cent. Sectoral growth was, however, rather mixed:

(*a*) The chief contributors to the boom were:

   (*i*) the automobile industry;
   (*ii*) the radio, telephone and electrical industries;
   (*iii*) the construction industry.

(*b*) Some sectors of industry were not so prosperous. Included in this category were:

   (*i*) agriculture (considered to be a main contributory factor in the collapse of the boom in 1929);
   (*ii*) coal-mining (especially bituminous mines);
   (*iii*) shipbuilding;
   (*iv*) cotton textiles (especially in New England);
   (*v*) railroads;
   (*vi*) flour milling;
   (*vii*) leather;
   (*viii*) lumber.

Uneven economic growth led to unemployment in these sectors of the economy. However, unemployment was not by any means spread among as high a proportion of the labour force as in Britain.

The effects of high growth rates among the "newer" industries were quite marked. Increasing demand for and production of motor vehicles, radios, telephones and houses had important multiplier effects benefiting a large number of other industries such as the steel, glass, rubber, cement industries, etc. Business confidence was high.

41. Flaws in the boom. The boom contained cancerous developments which brought about collapse in 1929. These were as follows:

(a) Demand for fairly expensive consumer goods was highly elastic. Much of the demand was new: replacement demand was low despite persuasive advertising and lavish credit facilities.

(b) Credit structures proved to be unsound: bank units were relatively small.

(c) Speculation in industrial shares, in real estate, etc., became widespread. This forced prices up to unrealistic levels, and induced a repatriation of American capital from abroad, which led to a fall-off in exports.

(d) Maldistribution of the national income had damaging effects:

(i) It left insufficient spending power in the hands of the mass consumers (60 per cent of the population received less than 24 per cent of the national income). Demand was based too much on credit.

(ii) Increasing profits left the bulk of the national income in the hands of a minority and led to: (1) demand for luxury goods – a highly elastic demand, easily reduced in unfavourable circumstances; (2) speculative activities of an unsound nature; and (3) increased capital investment leading to increased production and a build-up of inventories.

42. The collapse of the boom.

(a) The rate of growth of industries generating the boom (automobile, radio, electrical goods, telephone, construction, petroleum, etc.) slowed down and the rate of growth of investment declined. Business confidence began to ebb.

(b) The pyramided holding company structure of industrial groups proved susceptible to "reverse leverage" when growth slowed down—a deceleration effect was felt and accentuated by the corporations of this nature. Deflation began to replace inflation by October 1929.

(c) Share prices fell dramatically in October and November 1929 and were a symptom of a depression that became increasingly acute in 1932–3 and which continued in a less acute form until the Second World War reinvigorated demand for American products. Millions of dollars were wiped off share prices and those buying on credit were in difficulty; this in turn had serious repercussions for the credit structure and sent it toppling around the ears of investors and borrowers and, in turn, affecting industry.

43. The depression, 1929–32.

(a) *By mid 1933 share prices had fallen by 80 per cent*, prices of goods had been reduced by 30 per cent and employment had diminished by more than 25 per cent. The U.S.A. was in the grip of a deflationary spiral that had world-wide repercussions.

(b) *Industrial output fell to a nadir in 1932.* The coal, steel, automobile, textile, shipbuilding, electrical goods industries, etc., were compelled to cut back production and dismiss workers as demand fell and prices and profits diminished. Private capitalism seemed doomed as a satisfactory economic system.

(c) *To inject new confidence into business the state had to intervene*, Roosevelt's New Deal was intended to restore production to the 1929 level and to minimise unemployment. Industry was aided by the *National Industrial Recovery Act* and other measures.

PROGRESS TEST 9

1. Why was the growth of American industry so rapid? (**1–3**)
2. Discuss the consolidation of American firms:

   (a) before 1914; (**4–7**)
   (b) after 1918. (**8, 9**)

3. How was capital provided for American industry? (**10–15**)
4. Outline the development of:

   (a) the iron and steel industry; (**16–27**)
   (b) the coal industry; (**28–30**)
   (c) the automobile industry. (**31–35**)

5. Discuss the significance of the electrical industries in the interwar years. (**36–38**)

6. Why did some sectors of industry experience a boom in the 1920s? (**40–41**)

7. What caused the depression in industry from 1929 to 1932? (**42–43**)

PART FIVE

TRANSPORT

CHAPTER X

ECONOMIC EFFECTS OF TRANSPORT DEVELOPMENT

1. International trade.

(a) *Improved transport facilities* (railways, steamships, canals, roads and more efficient sailing ships) enabled wider trading links between the U.S.A. and Britain to be forged. As transport costs fell so trade tended to increase to the mutual advantage of both countries. The U.S.A. had an important outlet for surplus agricultural products and Britain was able to concentrate more and more on industrial production. The U.S.A. provided Britain with foodstuffs and raw materials and was an important market for British goods. Growing American industrialisation and the imposition of protective tariffs began to change the pattern of trade in the later part of the nineteenth century.

(b) *Greater international interdependence* developed as transport improved. There was a growth of regional and international specialisation, with an exploitation of the theory of comparative advantage. Multilateral settlements became more common as the international economy became more complex.

(c) *Raw materials of a bulky nature* could, for the first time in history, be transported quickly, at a reasonable cost, from areas of primary production to areas where secondary production was highly developed. World-wide markets were opened up for producers: competition increased.

2. Industrialisation.

(a) *A tremendous expansion of industrialisation* took place, especially in Britain, the U.S.A., Germany, France and Belgium. Britain lost her hegemony in the industrial world, partly because of the effects of innovations in transport.

(b) *The rate of technological progress* speeded up as competition became more acute among large-scale producers, who searched for:

119

(*i*) ways and means of reducing costs;

(*ii*) ways and means of joining forces to minimise competition.

(*c*) *Cartels, trade associations and the growth of huge firms* became possible on a wider basis. Sophisticated, complex organisations developed, controlled from a central point. Such control would have been impossible without the development of a speedy and relatively sure means of communication provided by railways, steamships and the telegraph.

3. The extension of the frontier areas.

(*a*) *New territories were opened up* for economic exploitation by the spread of the railway network in the U.S.A., Canada, South America, Russia, India, Australia, Africa, etc. The exploitation of relatively low-cost areas of agricultural production had profound effects on the fortunes of farmers during the last quarter of the nineteenth century.

(*b*) *Railway companies played a key role* in opening up and settling large areas of the U.S.A. Economic growth in these areas was speeded up or initiated by railway development.

4. Capital.

(*a*) *Railways*, because of their large capital requirements, facilitated a more rapid acceptance of limited liability as a basis of company organisation: this released for productive use large pools of "blind" capital.

(*b*) *Mobility of capital was increased*. Migration of capital from one country to another was associated with railway development to a large extent in the nineteenth century.

(*c*) A *cost-reducing effect* was felt because improved transport reduced the need for large stocks of goods to be held: less capital was tied up in the distribution pipeline. The whole cycle of distribution was speeded up: resources were used more efficiently.

5. Labour.

(*a*) *Mobility of labour* improved. Railways and steamships made long-distance migration relatively easy, within a country, internationally, and on an intercontinental basis.

(*b*) *Immigrants* streamed into the U.S.A. in the nineteenth and early twentieth centuries. Urban areas absorbed a great proportion of migrants.

6. State intervention.

(a) Government interference was necessary in both Britain and the U.S.A. to control railway companies:

(i) Freight charges were controlled.
(ii) Minimum safety standards were imposed.

(b) In Britain state interference was eventually greater than in the U.S.A., culminating in the nationalisation of transport under the Labour government in 1948. Railways came under state control in both countries in the First World War. They were returned to private enterprise after the war, but in Britain railways were again controlled by the state in the Second World War.

(c) In the nineteenth century huge land grants were made to American railway companies by the federal government to promote settlement of the interior. Assistance was given in raising capital by the sale of railway bonds. In Britain railways were essentially private enterprise concerns operating without state aid, but subject to state intervention to control development from time to time.

7. Railway development. The following table shows how railways expanded:

	Mileage of track ('000 miles)					
	1840	*1880*	*1900*	*1920*	*1940*	*1950*
U.S.A.	3	84	194	253	247	238
U.K.	1	18	22	20	20	20

Obviously, the great expansion of railways from 1840 to 1900 would have profound effects on the demand for iron (later on steel), timber, fuel (coal), building materials, furnishing materials for coaches, etc. In turn these firms associated with the production of such materials would demand more labour, materials, capital, etc. Railways, therefore, were important initiators of economic growth as construction enterprises, and, as transport enterprises, they enabled economic growth to flourish more easily.

8. The automobile and aircraft in the twentieth century.

(a) In the twentieth century the *motor vehicle* has become an important contributor to economic growth, just as the railways were in the nineteenth century. Profound changes in social and economic life have resulted from the mass production and mass use of motor vehicles.

(*b*) *Air travel* is growing in economic and social significance but, until 1940, the production and use of aircraft were of relatively minor economic significance.

PROGRESS TEST 10

1. How did transport development affect the international economy before 1914? (**1**)

2. To what extent was industrialisation facilitated by transport innovations? (**2**)

3. Why did governments intervene in transport affairs? (**6**)

4. Examine the accelerator effects of railway development. (**7**)

CHAPTER XI

TRANSPORT IN BRITAIN

CANALS

1. Economic effects in the early nineteenth century.

(a) *General distribution costs* were reduced. China clay, iron ore, coal, cotton and heavy or bulky goods could be carried quite cheaply.

(b) *The centre of England* (the manufacturing Midlands) was opened to economic expansion. Towns grew because:

  (i) food supplies could be brought in cheaply;

  (ii) raw materials could be obtained at relatively low costs;

  (iii) industry grew, partly because canals facilitated the obtaining of factors of production, and because goods could be transported to markets at reasonable cost.

(c) *The character of internal trade* began to change. Canals attracted trade away from rivers and packhorse routes. Commercial travellers, selling by name, became more common.

(d) *Employment opportunities* were created both in the construction and in the operation of canals. Valuable civil engineering experience was gained.

(e) *Investment opportunities* were created for private investors. The companies were essentially private enterprise in nature (except for Caledonian and Crinan Canals) and operated mainly on the joint-stock principle. (The Duke of Bridgewater's canals were privately owned.) Country bankers played a prominent part in the financing of canal development. A company framework was built up – a framework that could be adopted by the railway companies.

2. Decline of canals.

(a) *The profitability of canal companies varied enormously.* About £13 million was invested in eighty canals, with dividends averaging $5\frac{3}{4}$ per cent – but on nearly £4 million of this capital no dividend was paid. Canals in the south of England were markedly unsuccessful because they were constructed in areas where traffic was too light to produce economic returns.

(b) *Inadequacy of construction.* Banks were not strong enough to withstand pressures created by steam-driven vehicles. There was a lack of uniformity of depth and width. In urban areas often the widening of a canal was not feasible and modernisation was impossible.

(c) *Canal companies did not act as carriers* and failed to organise traffic on the canals.

(d) *Railways could be built in more convenient places* than canals and provided more certain, speedy means of transport. They were less affected by the variances of the weather.

(e) *Development of steam coasters.* Steam-driven coasters attracted bulk traffic away from canals.

(f) *Strategic sections of canals* were purchased by railway companies, effectively killing off competition from such canals.

(g) *No rationalisation.* The canal system lacked any large-scale rationalisation during construction. Canals were of differing sizes, locks varied in type and size and tolls had to be paid at too frequent stages. All these factors lessened the competitive value of the canals.

RAILWAYS

3. The development of railways.

(a) *Early successes.* The Carlisle–Newcastle, Stockton–Darlington and Liverpool–Manchester railways created an image of success and profitability in the minds of the public. Between 1825 and 1840 Parliament passed 98 Railway Acts. By June 1843 1,900 miles (3,040 km) of railway were open to the public.

(b) *Railway mania.* Growing economic prosperity in the mid 1840s was accompanied by a great interest in railway promotion. The years 1844–6 saw a tremendous increase in railway promotion and development. By 1848 12,000 miles (19,200 km) of railway development had been authorised and 5,000 miles (8,000 km) were open to the public.

(c) *Railway traffic.* By 1850 railways were more important as a means of national transport for both passengers and goods than either canals or roads.

(d) *Amalgamation.* The 1840s were years of great activity in railway amalgamation: George Hudson was a notable leader in railway consolidation. Twenty-six Acts of Parliament concerning railway amalgamations were passed between 1844 and 1846. The Midland Railway and the London and North-eastern Railway were created by amalgamations of competing and complementary companies.

(e) *1848–1900*. By 1900 railway mileage totalled 18,680 miles (29,888 km), much of this consisting of an increase in the capacity of established routes and the opening of branch lines. Amalgamations continued:

    (i) 1854. The North-eastern Company was established.

    (ii) 1862. The Great Eastern Company was established.

    (iii) 1863. The Great Western Railway Company took over the West Midland and South-west Railways.

    (iv) 1911. Eleven groups controlled British railways, but more than 1,000 Acts of Parliament had been passed to create railway companies.

(f) *Uniformity of gauge*. This was finally achieved in 1892, although the Act of 1846 stated that the narrow gauge of 4 ft 8½ in. (143·51 cm) should be used. The Great Western adhered to the broad gauge until 1892.

4. The nature of railway development.

(a) *Private enterprise:*

    (i) Railway companies in Britain, unlike many in the U.S.A., did not receive initial state assistance. Companies were joint-stock enterprises with limited liability: they illustrated clearly that the limited liability form of organisation could be adopted successfully but that safeguards for investors were needed. Capital was raised by the flotation of shares: existing shareholders in railway companies were usually given the first option on new shares issued, and a proportion of stock was often reserved for landowners and inhabitants of towns *en route*. Liverpool businessmen were active investors in the early railways.

    (ii) Estimated costs per route mile in 1855 were nearly £40,000 and by 1904 £66,000. Capital requirements, therefore, were large. Provincial stock exchanges were set up to deal in railway shares.

(b) *Rationalisation:*

    (i) Dalhousie drew up a plan for an integrated system in 1844, but initially companies were relatively small with short stretches of railway track under their control. Rationalisation was achieved in piecemeal fashion through amalgamations in the 1840s and between 1850 and 1921; nationalisation took place in 1948 (*ii*) Parliament usually imposed conditions on companies when passing incorporating Acts to prevent local transport monopolies arising. Amalgamations aroused greater public fears of monopoly; legislation was necessary to curb the growing power of railway groups. General railway legislation enabled Parliament to control freight charges, passenger fares, safety measures, etc.

(*c*) *Private rolling stock.* A feature of railways was the existence of privately owned rolling stock. Many companies owned wagons and trucks: the practice continued until 1948.

5. The state and railway development.

(*a*) 1838. The *Railways (Conveyance of Mails) Act* was passed.

(*b*) 1839. A *Select Committee* reported that "every railway should be under one system of management under one superintending authority." Railway companies needed to act as carriers to prevent chaos on railway lines.

(*c*) 1840. Lord Seymour's *Railway Regulations Act* gave the Board of Trade general powers to supervise railways and to require returns of freight rates, passenger fares, the volume of traffic and accidents. (The Act was amended in 1842.)

A Board of Trade licence was necessary before a passenger line could be opened.

(*d*) 1844. *Gladstone's Act* was passed:

(*i*) A parliamentary fare of 1*d.* per mile was introduced for third-class passengers.

(*ii*) A provision for the purchase of railways after a period of twenty-one years was included.

(*e*) 1844. The *Railway Department of the Board of Trade* was established under Lord Dalhousie to examine railway bills. It was defunct after one year.

(*f*) 1846. A *Railway Commission* was created to supervise railways: it lasted six years. The narrow gauge was adopted.

(*g*) 1854. *Cardwell's Act* was passed to end undue preferential rates and to compel railway companies to provide facilities for through traffic.

(*h*) 1868. The *Regulation of Railways Act* compelled companies to keep accounts in a specified form.

(*i*) 1873. The *Railway and Canal Traffic Act* saw the beginning of more intense state control of railways. The Railway Commission was appointed for five years with the following brief:

(*i*) To examine proposed amalgamations and working agreements.

(*ii*) To examine proposals of railway companies to purchase canals.

(*iii*) To hear complaints about preferences and to decide on the reasonableness of rates.

(*iv*) To decide on proper terminal charges.

(*j*) 1888. The *Railway and Canal Traffic Act* was passed:

(*i*) It made the Railway Commission permanent and enlarged its powers.

(*ii*) All rates were to be printed in a rates book, which was to be open for public inspection.

(*k*) 1894. The *Railway and Canal Commission* was virtually given control of railway rates that exceeded the 1892 level. This tended to make railway charges inflexible. Competition turned from price to service and railway companies negotiated working agreements with each other where amalgamation was precluded by Parliament.

(*l*) 1913. The *Railway and Canal Traffic Act* was passed. Railway companies were permitted to raise rates above the 1892 level.

(*m*) *Nationalisation.* By 1914 it was mooted by many that railways should be owned and controlled by the nation. Large railway groups, it was alleged, operated at times against true national interests, and efforts to control railways were thought to have failed.

6. The state and railways after 1910.

(*a*) *The First World War and railways:*

(*i*) During the war the railways were operated as a unified system under the control of a Railway Executive Committee consisting of the general managers of the chief companies.

(*ii*) Earnings were guaranteed at the 1913 level and inter-company competition ended: impressive economies in operation were achieved.

(*b*) *Select Committee on Transport* (1918). The Report of the Committee recommended ultimate unification of the railways.

(*c*) *Railway Act*, 1921:

(*i*) Railways were reorganised into four groups under the Railway Amalgamation Tribunal so as to achieve economies of scale and standardisation. Regional monopolies were granted, but there were overlaps.

(*ii*) A system of standard charges based on mileage was introduced. The Railway Rates Tribunal had to be consulted if companies wished to impose different charges from those laid down.

(d) *Royal Commission on Transport* (1930). This investigated the position of railway companies to ascertain the extent of the success of the 1921 Act and concluded that the economy and efficiency hoped for in 1921 had not been achieved. Receipts fell from 1923 and the standard net revenue of £51 million was never reached in the inter-war period. The inflexible system of fixing freight rates and fares proved too rigid in view of the increasing competition from road transport. It is estimated that about a quarter of the railways' capital went without dividend by 1932 (total capital was about £1,120 million).

(e) *The Railways (Agreement) Act*, 1935. This Act provided for a loan to railway companies of £26·5 million for modernisation, but difficulties continued because of lower-cost road transport, and the fall-off in heavy freight traffic. (In 1928 railway companies were empowered to link themselves with other transport undertakings.)

7. Inter-war difficulties. Financial difficulties were experienced by railway companies in the period 1921–39 because of a fall-off in freight and passenger traffic. Freight carried by railways diminished: annual coal traffic diminished by about 30 million tons between 1929 and 1936. The loss of freight and passengers was due to various factors:

(a) There was a decline in the volume of exports as compared with 1913. The staple industries were particularly affected by this: iron and steel, cotton and coal freight fell below pre-war levels. This freight was bulky in relation to its value and its loss was a blow to the railways, which were particularly suitable for the carriage of such goods.

(b) Emerging industries such as the radio, domestic appliance and motor vehicle industries found road transport in many cases *more economic* and *more flexible* than railroads.

(c) Railway companies had to *publish their rates* and provide the nation with a *wide service*. Road hauliers were under no such compulsion: they could cut rates and choose areas of operation.

(d) *Omnibus companies mushroomed* after the First World War and captured passengers from railways, especially over short and medium distance routes. In 1920 railways carried more than 1,500 million passengers, but in 1930 the number had declined to 844 million.

8. Railways during and after the Second World War.

(a) In the Second World War railways were again placed under *state control*. The companies were guaranteed an income of £43 million. Maintenance was neglected, and in 1945 a terrific maintenance lag had to be made up, necessitating large capital schemes for reconstruction.

(*b*) *Nationalisation* was one means of providing a source of finance for reconstruction. Railways were considered as one part of a national system of transport under state ownership.

(*c*) The *Transport Act, 1947*, contained certain important provisions:

(*i*) *The British Transport Commission* was set up: (1) to acquire control of railway companies and their subsidiaries; (2) to acquire control of road haulage; (3) to control harbours and inland waterways.

(*ii*) *The Transport Tribunal* was set up to approve rates and fares to be charged.

(*iii*) *Railways were nationalised* on 1st January 1948, and a programme of reconstruction was put into effect to restore railways to pre-war efficiency. In addition work was commenced on the formulation of longer-term plans for the modernisation of railways to secure economies and greater efficiency to meet post-war needs.

(*d*) In 1955 a *fifteen-year plan* for modernisation and the rationalisation of British railways was launched, but by 1960 competition from other forms of transport was increasing. Railways were subject to reappraisal, and in 1962 the *Transport Act* was passed and the British Railways Board was created with Dr Beeching as chairman: he was responsible for the formulation of plans to rationalise the railway system by pruning uneconomic lines.

SHIPPING

9. Technological advances.

(*a*) *Metallurgical improvements* enabled metal ships to be constructed and used safely on the high seas. Important economies were brought within reach of shipowners because the size of ships was no longer restricted within the limits imposed by the use of timber in shipbuilding.

(*b*) Britain was in a good position to obtain economic advantage from the developments in the application of metallurgical science to shipbuilding in the second half of the nineteenth century. She had adequate supplies of *coal* for smelting iron ore and then producing steel, and had access to supplies of *iron ore*. These factors, linked with her *expansion of foreign trade*, facilitated the development of shipbuilding in Glasgow, Birkenhead, etc.

(c) The development of the *compound steam engine*, the *triple* and then *quadruple expansion engines*, and the *steam turbine* increased the speed, reliability and economy of steamships in the nineteenth and twentieth centuries and were of crucial importance in the build-up of Britain's mercantile marine. Cost-reducing innovations such as these enabled British shippers to compete with shipowners from lower-wage countries such as Japan, Greece and Norway.

10. British shipping, 1850–1914. Mercantile supremacy was enjoyed by Britain's shipowners before the First World War, based on industrialisation (and the relatively early development of large-scale steel production), the coal trade and an expansion of other overseas trade, a rapidly increasing population, and an ability to harness supplies of capital for shipbuilding and commerce

(a) *The expansion of coal exports* was of great significance, since it eliminated or minimised the need for ships to make outward journeys in ballast and, therefore, helped keep down freight charges.

(b) *The relatively early industrialisation* was a key factor in the growth of Britain's shipbuilding and mercantile marine between 1850 and 1914:

(i) Whilst wooden sailing ships operated at economic advantage over metal steamships, British shipbuilders were handicapped in international competition by the relatively high cost of timber in Britain. Indigenous supplies of timber were scarce in Britain, but plentiful and cheaper in the U.S.A.

(ii) The success of the metal steamship put Britain's shipbuilders and shipowners in an advantageous position because of the existance of relatively sophisticated iron and steel producing plants. Comparative costs turned in favour of shipping entrepreneurs in Britain.

(iii) Technological advances enabled owners of Britain's tramp steamers to operate so effectively and competitively that earnings of shipping abroad grew by almost 25 per cent between 1871–5 and 1896–1900.

(c) *The increase in population* was an important factor in the development of Britain's shipping from three aspects:

(i) It provided an increasing pool of labour for shipbuilders and shipowners and thereby helped keep labour costs relatively low.

(*ii*) The increasing population could be fed only by increasing imports of foodstuffs from abroad, paid for in part by increasing exports of manufactures, and shipowners benefited from the resultant increase in trade.

(*iii*) Passenger line services were helped in their development by the fact that between 1865 and 1894 Britain was the main source of emigrants to the U.S.A. An average of 119,000 people a year emigrated from Britain to the New World.

11. British and American mercantile tonnage. In 1850 the British mercantile fleet totalled approximately 3·8 million tons—more than 90 per cent being sailing tonnage. In 1912 total tonnage was nearly 12 million tons, of which more than 90 per cent was steam tonnage. The American fleet was 1·6 million tons in 1850 and still about the same in 1912 and only 60 per cent was steam tonnage. There was a massive build-up of the American fleet towards the end of the First World War.

12. Shipping conferences.

(*a*) *Intense competition* among shipping concerns developed between 1870 and 1914 as the world's shipping capacity increased. The cost structure of shipping is such that rates can be forced down to a level where prime costs only are covered. To minimise such competition conferences were set up among the cargo liner services using the deferred rebate weapon to retain customers within the shipping "rings." Examples of these conferences were as follows:

(*i*) 1875. Calcutta Conference.
(*ii*) 1879. China Trade Conference.
(*iii*) 1884. Australia Trade Conference.
(*iv*) 1886. West Africa Conference.
(*v*) 1895. South Africa Conference and Brazil Conference.
(*vi*) 1904. South America Conference.
(*vii*) 1908. Atlantic and Mediterranean Conferences.

(*b*) *Tramp steamship owners* usually operated outside the shipping "rings" because of the irregular nature of their contracts and the great variety of conditions under which they operated. Conference rates could be enforced only where regular transport was wanted at particular times.

(*c*) *The Royal Commission on Shipping Rings* (1908) considered that shipping conferences were beneficial to shippers and customers despite the monopolistic threats they contained.

13. Shipping and the state.

(*a*) *Government mail contracts* were an important factor in the early stages of the growth of steamship lines. Postal subventions helped subsidise steamship companies, filling the gap between total costs and the normal commercial revenue obtainable in the 1840s and 1850s, thus helping companies to grow beyond the stage of infancy to one in which they could thrive independently. The Cunard, P. & O., Wilson, Allan and Union Castle lines were assisted in this way.

(*b*) The state intervened to impose *minimum safety standards* under the 1875, 1876 and 1894 Acts, which laid down loading limitations and prescribed minimum requirements for rescue apparatus on ships.

14. The First World War and its effects.

(*a*) *Shipping losses* were extremely heavy, some 9 million tons of shipping being destroyed, resulting in a decline in total tonnage by 2·6 million tons between 1914 and 1919, at a time when the American merchant fleet was growing from 2 million tons to 9·8 million tons.

(*b*) *Pressure on shipping* for war-time purposes, and *the submarine menace*, brought about a suspension of long-haul liner services, which prompted a more speedy development of foreign shipping, leading to a loss of entrepôt trade for Britain's shipping lines.

(*c*) The effect of the war on *shipowners' profits* varied according to whether Blue Book rates (agreed between shipowners and the government in 1915) or free-trade market rates were payable. Blue Book rates were payable for requisitioned shipping space and for government contracts, and whilst these rates were quite remunerative in 1915 and 1916 they were relatively inflexible at a time when free-market rates were rising rapidly. Profits made during the war could have ensured adequate fleet replacement had they been conserved. Speculative entry into shipping was encouraged by the high profit potential, and this had certain effects:

(*i*) It helped force up ship prices.

(*ii*) Shipping shares were sold at inflated prices and investors suffered large capital losses when share prices fell in 1920.

(*iii*) Amalgamations among companies occurred. Liner companies were attracted by the profitability of tramp freighting and, to some degree, this made quick readjustment difficult in the inter-war period.

15. The inter-war years. The boom in shipping continued in 1919 and the early part of 1920. There was considerable speculation in shipping shares. A break in freight rates and shipping prices occurred in 1920 as the world's ports released congested shipping. The relationship between the quantities of world trade and shipping space available changed: the latter became a less scarce factor. Difficulties continued after 1920 because of the following:

(a) British exports, in volume, failed to reach the level of 1913 and, in particular, except in 1923, coal exports were well below the pre-war figures.

(b) The loss of coal exports occurred mainly in the long-haul coal-trade routes and this loss was a serious cost disadvantage for British shipowners, especially the tramp owners.

(c) Diesel propulsion was adopted as a more efficient means of power unit in modern motorised ships. Scandinavian and Dutch shipowners, in particular, modernised their fleets and wrestled trade from British shippers unwilling or unable to modernise. The raising of new capital for fleet modernisation was difficult because of the experience of investors in 1920–1.

(d) Subsidies and protective policies aided some foreign rivals to take trade away from British shippers.

(e) British shipowners tended to be too inflexible and were not aggressive enough in seeking new trades within which to operate. There was too great a reliance on traditional markets and trades and insufficient energy was displayed in endeavouring to provide the right types of ships for particular trades, e.g. the fruit trade.

(f) Between 1925 and 1931 the pound sterling was on the gold standard at an exchange rate which Keynes considered was 10 per cent too high and which, therefore, handicapped British tramp owners in world markets.

(g) American restrictions on immigration cut down the flow of migrants who had formed an important part of the passenger trade in the pre-war period.

(h) The great depression beginning in 1929 reduced world trade and thereby affected freight rates, whilst nations became more economically nationalistic in commercial policies. Trade tended to become more bilateral and less multilateral, to the disadvantage of the tramp owners.

16. The Second World War and its effects.

(a) *Freight rates* were controlled during the war and, in 1940, a requisitioning scheme was imposed. All British-owned vessels were taken under charter by the government, with liner companies being responsible for the management of vessels. The government ensured freight rates and compensation payments for ships lost were fair and not excessive: the anomalous situations of the First World War were not repeated.

(*b*) *Losses* amounted to about 11·63 million tons but the British fleet in 1946 was only 3·6 million tons less in size than in 1939, having diminished by about 21 per cent. The American fleet grew from 8·7 million tons to 40·9 million tons—an increase of 370 per cent.

(*c*) After 1945 shipowners were cautious about ship replacement and about fleet expansion because of fears of a price collapse, despite the profits earned by liner and tramp owners. Opportunities in the post-war years were lost through this conservatism: British shipping in the period 1945–58 could, with profit, have expanded fleets to a much greater extent if it had been willing to break away from traditional self-finance.

ROAD TRANSPORT

17. Effect of railway competition in the nineteenth century. Long-distance road transport was largely killed off by the growth of railways but short-distance transport to and from railway stations thrived. Turnpike trusts and long-distance coaching were not competitive with railways: horse-drawn carts, cabs, omnibuses and trams increased until the appearance of efficient mechanical road transport.

18. The motor vehicle.

(*a*) *Passenger transport* increased and, in the inter-war period, captured a considerable proportion of railway trade. Intensive competition among omnibus companies induced state intervention, firstly in London and secondly in the provinces to control the development of omnibus services:

(*i*) The *London Traffic Act*, 1924, gave the Minister of Transport control over companies operating in London.

(*ii*) The *Road Traffic Act*, 1930, created thirteen traffic areas under the control of Traffic Commissioners who were responsible for the issuing of public transport licences and could regulate conditions of service.

(*iii*) The *London Passenger Transport Act*, 1933, created the

London Passenger Transport Board, which took over all omnibus passenger transport in London.

The effect of the Acts was to eliminate wasteful competition in public passenger transport.

(b) Road goods transport grew in the inter-war period from a minor industry into a major one. By 1939 nearly 500,000 vehicles were in operation on the roads of England.

(c) The *Road and Rail Traffic Act*, 1933, imposed control on the industry in respect of licences, speed limits, conditions of vehicles and conditions of work of transport drivers. Goods transport firms were able to capture much of the railways' traffic because they were able to operate more flexibly: they were willing and able to adapt themselves to customers' needs, they could operate from door to door, and there was no legal stipulation that rates should be published such as applied to railway companies.

(d) The *Transport Act*, 1947, set up the British Transport Commission, which, as one of its tasks, was to acquire control of road haulage.

19. The motor vehicle industry. The production of motor vehicles was an important growth point in the economy both during the inter-war period and after 1945. In 1938 there were some 3 million road vehicles in Britain, increasing to 10·5 million by 1962. Between 1933 and 1938 the export of vehicles increased tremendously, and the industry formed a key part of Britain's export drive after 1945 and since then the industry has had a dominant place in the economy.

AIR TRANSPORT

20. Growth of air services. The build-up of international and internal air services began in the inter-war period, but not until the period after the Second World War did air transport become of considerable economic importance.

	Aircraft mileage	Passengers	Tons of cargo
1919	104,000	870	30
1937	10,770,000	244,000	4,000
1960	109,000,000	6,000,000	240,000

PROGRESS TEST 11

1. Discuss British railway development in the nineteenth century. (**3, 4**)

2. Account for government intervention in railway affairs. (**5–8**)

3. Why were railway companies in difficulty after 1920? (**7**)

4. Why were the shipping conference rings formed? (**12**)

5. Comment on economic trends in shipbuilding and shipping after 1900. (**12, 14, 15**)

6. What were the effects of the First World War on shipping? (**14**)

7. Assess the importance of road transport in the inter-war years. (**18, 19**)

CHAPTER XII

TRANSPORT IN THE U.S.A.

TRANSPORT BEFORE 1860

1. Road transport. Turnpike roads were constructed by both private corporations and public trusts: generally the former were found in the North and the latter in the South. Private corporations built and maintained roads under the spur of the profit motive. Approximately 3,750 miles (6,000 km) of road were built by private corporations in New England by 1838. Most corporations were small and had capital less than $100,000 each. The capital was provided by private investors, especially merchants and bankers, by local authorities and by state governments. Sufficient capital was raised to provide roads linking all the major cities in the North and the East by 1820. State investment was greater in the West and the South than in the North and the East.

(*a*) Turnpikes *failed to provide cheap transportation* over long distances and, usually, were unprofitable. Costs frequently exceeded income from tolls and, therefore, even prime or variable costs were not covered, whilst nothing was contributed to fixed or supplementary costs.

Companies failed because of:

   (*i*) higher construction costs than anticipated;
   (*ii*) high costs of operation and maintenance;
   (*iii*) toll evasion by road users;
   (*iv*) competition from canals, navigable rivers, coastal shipping and railways.

By 1825 turnpike companies were a spent force.

(*b*) *Constitutional difficulties* prevented the federal government from playing a major role in road development, although the National Pike was constructed at a cost of $7 million, and aid was given for the construction of national roads.

2. Canals.

(*a*) Canals were constructed in two ways:

(*i*) By private corporations (sometimes with state aid).

(*ii*) As public works under state governments.

(*b*) *The Erie Canal:*

(*i*) The great era of canal construction came with the opening of the Erie Canal in 1825. By 1850 3,700 miles (5,920 km) of canals had been built.

(*ii*) Capital for the Erie Canal was provided by the State of New York, which financed construction by loans obtained mainly from Britain.

The Canal had a great impact. (1) Travelling time from Buffalo to New York was reduced from twenty days to six days. (2) Freight charges fell drastically. (3) Land values in New York State were enhanced. (4) The rate of growth of the port of New York speeded up. (5) Trade was pulled west to east: the pull of the South via the Mississippi river system diminished relatively. (6) Industrialisation of the North-east increased. (7) The success of the Erie Canal attracted further capital from abroad for canal development. (By 1882, when tolls were abolished, $120 million had been collected on the Erie Canal, which cost about $7 million to build.)

(*c*) *Other canal development followed rapidly.* Many companies relied on aid from state governments and, to a lesser extent, from the federal government. In 1820–4 $13 million worth of state bonds were issued to raise capital for canals and in 1835–7 $60 million worth of bonds were issued. State debts in 1840 totalled $200 million. Canals were built by corporations, with state aid, in New England, Virginia, Maryland, etc. The federal government made significant contributions to the Chesapeake and Ohio, the Chesapeake and Delaware and the Dismal Swamp Canals, besides making land grants for canal developments in Ohio, Michigan, Indiana, Illinois and Wisconsin.

(*d*) By 1840 *the end of canal building* had almost come. The financial crises of 1837 and 1839 brought some states to near bankruptcy: interest on borrowed capital was not paid. The states of Mississippi, Louisiana, Maryland, Pennsylvania, Indiana and Michigan repudiated debts. Foreign investors were deterred by these experiences.

3. Other navigable waterways.

(*a*) *Rivers, the Great Lakes and coastal waters* provided economical routes for the long-distance transportation of bulky goods. Steam-engine development helped revolutionise transport by water. From 1830 to 1850 the river steamboat was the most important type of internal transport – turnpikes and canals were complementary rather than competitive. Important areas of steamboat operation were:

    (*i*) routes radiating from New York;
    (*ii*) the Great Lake area;
    (*iii*) the West, where freight tended to be more important than
           passenger traffic.

(*b*) *Capital for steamboat companies* was raised mainly from local sources – merchants, manufacturers, farmers, insurance companies, etc.

(*c*) *Capital for improvement of river navigation* was provided by private navigation companies and by state governments. Federal land grants were made to assist in river improvements.

(*d*) *Federal and state regulations* were introduced to control the construction and operation of steamboats. The *Steamboat Act*, 1852, provided for effective inspection of steamboats to ensure compliance with regulations.

4. Railroads.

(*a*) There was a *dramatic growth* in railways in the U.S.A. The steam railroad provided one solution to the problem of developing the interior of the country. By 1860 30,000 miles (48,000 km) of track had been laid. American railway companies were able to expand their activities more easily than their English counterparts because of the following factors:

    (*i*) Urgent demands for better inland transport.
    (*ii*) Lack of political restrictions.
    (*iii*) Cheapness of land.
    (*iv*) Less well-entrenched opposition from vested interests.

(*b*) *Capital needs* were enormous and involved a high risk factor. American railroad companies, therefore, were organised as corporations with limited liability under state charters, which frequently contained great privileges for the companies in relation to taxation, banking and the minimising of competition.

(*c*) *Private capital* was invested by people in the locality of railway development – merchants, manufacturers, farmers, etc., who had something to gain from transport improvements, but local capital had to be supplemented by capital drawn from the eastern seaboard cities. Stock exchanges in Boston and New York were important intermediaries in raising capital.

(*d*) *Foreign capital* was of significance. By 1870 more than $240 million worth of railway stocks and bonds were in foreign hands. British capital was freely drawn upon.

(*e*) *State governments* built and owned railways in Georgia, Virginia, Michigan, Indiana and Illinois. State governments also facilitated railway development by supplying finance and credit for construction.

(*f*) *Federal assistance* for railroad companies took three forms:

    (*i*) Railway surveys were made at federal expense.

    (*ii*) The imports of iron rails, etc., for railroads were admitted at preferential rates of duty.

    (*iii*) Land grants were made to railroad companies under Acts of the 1850s. These grants were of significance because: (1) they freed companies from the expense of acquiring land by purchase; (2) capital could be raised on the security of the land granted to the companies; (3) the federal government benefited in that railway companies agreed to convey federal troops and property free of charge, and to convey mail at rates fixed by Congress.

(*g*) Railways reinforced the *west-to-east pull* on trade that was exerted by the Erie Canal.

RAILWAYS, 1865–1914

5. The economic significance of railroads. The established popular view of the great relative economic significance of railroads in the U.S.A. has been queried, but, whilst there was a lack of planning and rationalisation about railroad development in the U.S.A., which caused considerable economic waste, undoubtedly railroads were important innovations initiating rapid economic growth. (As in Britain there was a risk of development of unnecessary competing lines to areas where traffic did not warrant such provision – especially in the Prairies.) Railroads were not solely the cause of rapid economic growth, but they helped supply relatively cheap inland transport that was a condition for the economic growth which occurred. They helped boost demand for iron, steel, timber, textiles, leather, etc., and they provided employment opportunities for thousands of people. As creators of demand for materials, as creators of employment, and as transformers of the capital market, railroads were of vast economic significance in the U.S.A. in the nineteenth century.

(*a*) *Railroad construction* had a somewhat discontinuous rapid rate of growth to about 1890. Demand for iron reached a peak in the 1860s and that for steel in the 1880–5 period. High labour demand and the need to promote settlement of the West induced railway companies to organise immigration schemes which helped swell the growth of the population, helping thereby to create a market-orientated economy, making mass production of consumer goods possible and profitable. Between 1860 and 1900 railroad mileage increased from 30,000 miles (48,000 km) to about 200,000 miles (320,000 km): in 1914 it was 250,000 miles (400,000 km.).

(b) *Urban growth was stimulated*, especially at the seaport terminal points such as Boston, Philadelphia and New York, at junction towns such as Akron and South Bend, and at inland terminals such as Chicago, Buffalo, Cincinnati and St Louis. At these centres manufacturing, processing plants, railway works, etc., thrived.

(c) *Transport costs fell.* By 1895 freight rates were about 0·84 cent per ton-mile for bulky goods on long-distance routes (in 1867 the figure was 1·90 cents). In 1870 to travel from New York to St Louis took forty-two hours and cost $46: in 1900 it took twenty-nine hours and cost $37. Passenger fares per route mile fell by about half.

(d) *Innovations in financial enterprises* developed as a result of railroad finance. There was a rise of specialist firms in investment banking, acting as trustees and registration agencies for securities. Holding companies became a familiar part of the American economic scene and eventually financiers (*e.g.* J. P. Morgan) began to exert control over railway companies. By 1910 some $18 billion was invested in American railroads.

6. Land grants. Public policy towards railroads in the second half of the nineteenth century still arouses controversy. Congress is alleged by some historians to have been excessively generous in its land grants and financial aid for railway companies, but one school of thought argues that:

(a) the amount of land taken over by railroad companies has been exaggerated, although it was still a vast amount;

(b) the federal government reaped enormous economic benefit from the free carriage of federal troops and property and from mail contracts;

(c) loans to railroad companies totalled $64 million, but by 1900, in interest and repayment, nearly treble this amount was returned;

(*d*) land grants to railroads were the best means of securing settlement of land in the West and of securing construction of a transportation network.

Estimates of the total land grants from federal, state and local authorities vary from 216 million acres (86·4 m. ha) down to 131 million acres (52·4 m. ha), but whichever is correct a vast area of land was handed over.

7. Finance for railroads.

(*a*) *Capital sources* were as outlined previously (*see* **4** above) but development of railroads in the West raised peculiar problems. Capital could not be raised locally, since the railroads were developed ahead of settlement. Finance was raised by promoters, who formed construction companies to which railroad stock was given in return for the construction of the lines. The construction companies disposed of stock to the public and hired contractors to build the railways. "Watering" of the stock often occurred and handsome profits were made by the promoters. The most notorious example was that of the "Credit Mobilier Co." which made a profit of $50 million on the construction of the Union Pacific line.

(*b*) *Foreign capital* continued to be significant until 1914–18. Railroad securities found a good market in Europe and especially in Britain, but during the First World War many foreign holders of stock disposed of their holdings to the U.S.A. In 1914 foreign-owned securities were valued at less than one-fifth of the total value of railroads: private domestic capital was the most important source of finance.

8. Railroad consolidation.

(*a*) *Combination* of railroad companies took place, very much as in Britain:

    (*i*) on an end-on basis giving a group control over continuous stretches of track thereby creating greater efficiency;

    (*ii*) to consolidate operations in a particular area—a combination of competing groups to eliminate competition.

(*b*) *Fierce competition* accompanied by a forcing downwards of freight and passenger rates made consolidation necessary, since pooling arrangements worked inefficiently and then were made illegal by the *Interstate Commerce Act*. By 1906 seven groups controlled about 155,000 miles (248,000 km) of railway out of a total mileage of approximately 228,000 (364,800 km) and controlled more than 80 per cent of railway revenue. Bankers were prominent in these groups and built up interlocking directorates in numerous interests. Reorganisation often gave promoters a chance to "water" stock and some railways were, as a consequence, heavily burdened with debt.

(c) *Banking houses*, such as J. P. Morgan & Co., used the opportunity created by the panic of 1893 to acquire control of railway companies, when more than 40,000 miles (64,000 km) of railroad were in receivers' hands.

9. Regulation of railroads.

(a) Discriminatory railway charges caused much unrest, especially among farmers, in the 1860s and 1870s. Pressure was exerted by the *Granger Movement*, and the so-called *"Granger Laws"* were passed in the 1870s:

(i) The State of Illinois passed Regulatory Acts in 1869 and 1871 controlling railway freight rates, passenger fares and regulating warehouses.

(ii) In 1871 Minnesota followed suit by fixing freight charges and passenger fares, and in 1874 Iowa and Wisconsin passed similar Acts.

(b) In the *"Granger Cases"* (*Munn* v. *Illinois* (1867) and *Peik* v. *Chicago & North-western Railway Co.* (1876)), the power of states to regulate railroad companies was upheld, but in 1886 and 1889 the Supreme Court reversed these rulings:

(i) The Supreme Court held that they involved interstate commerce and states had no jurisdiction beyond their own boundaries.

(ii) The court also held that reasonableness of rates was a judicial and not a state legislative matter.

(c) *Federal regulation* had to be imposed following these rulings:

(i) 1887. The *Interstate Commerce Act* forbade railway companies to charge at discriminatory rates, prohibited pooling and ordered publication of fares and rates. An Interstate Commerce Commission was set up to administer the Act, but it was given power only to investigate and to institute court proceedings, and railway companies were able to thwart its actions.

(ii) 1890. The *Sherman Anti-Trust Act* declared illegal combinations of firms aimed at monopolising or restraining trade. However, it did little to restrain railroad consolidation and monopoly practices.

(iii) 1903. The *Elkins Act* attempted to end the rebate system.

(iv) 1906. The *Hepburn Act* gave teeth to the Interstate Commerce Commission by empowering it to determine railway rates and fares.

(*v*) 1910. The *Mann–Elkins Act* enlarged the powers of the Interstate Commerce Commission and set up a special Commerce Court to hear railroad cases.

10. Technological developments.

(*a*) Metallurgical improvements, developments in boiler technology, automatic brakes, automatic and flexible couplings, improved signalling devices, etc., enabled locomotives of greater power to pull larger trains. Freight cars between 1870 and 1920 increased in size from a capacity of 10 tons to a capacity of 100 tons. Aggregate improvements in technology had significant cost-reducing effects.

(*b*) *Electric railways* were developed in the 1880s and became important forms of transport in cities and, by 1920, became of significance in inter-urban transport.

RAILWAYS, 1914–50

11. Difficulties.

(*a*) Railways were taken over by the federal government in 1917 and returned to ownership in 1920 under the *Transportation Act* of that year. Economic health was to be maintained by financial assistance from the federal government and by further consolidation and rationalisation, but in the 1920s passenger traffic fell by 40 per cent, and fell even further in the 1930s.

(*b*) Loans from the *Reconstruction Finance Corporation* were needed to save some lines from closure in 1932. In 1933 the *Emergency Transportation Act* was passed and this provided for the co-ordination of railroads through a Federal Railroad Co-ordinator, but difficulties continued because of:

   (*i*) competition from road transport;
   (*ii*) competition from pipelines;
   (*iii*) the fall-off in trade – foreign and domestic;
   (*iv*) relatively high costs of operation on railroads;
   (*v*) increasing competition from air transport for long-distance passenger services.

(*c*) In the Second World War the railroads were not put under federal administration, but were subject to greater regulation. A tremendous increase in passengers and freight was handled successfully. After the end of the war railways again had difficulty in meeting competition from road, air and pipeline transport, partly because cost reductions gained by technological advances were offset by changes in the wage structure of the labour force, making it difficult to pass on benefits to customers.

SHIPPING

12. The American mercantile marine. Coastal and internal trade took up the bulk of shipping, except for the years 1919–21 and the Second World War. Between 1865 and 1915 the mercantile marine increased from about 3 million tons to 8·4 million tons: in 1921 it totalled 18 million tons. There was a decline in the 1920s and 1930s to 14·5 million tons in 1939, but the Second World War boosted the size of the fleet to a massive total of more than 30 million tons.

13. Internal trade and coastal shipping. In the 1850s inland and coastal water routes carried the bulk of American trade, but the development of railways and, in the twentieth century, road transport and air transport eroded the position of water transport in a relative sense.

(a) Coastal traffic grew to 1914, but total inland trade grew at a much faster rate. By 1930 coastwise trade was about 9·3 per cent of railroad traffic in volume.

(b) The transportation of passengers and general merchandise was affected most of all. The transport of oil, grain, ore, coal, molasses, sulphur, etc., expanded, especially along the Great Lakes routes. Canal construction linked the Great Lakes and, in 1914, the Panama Canal linked the Atlantic and Pacific Oceans, but railways dominated inland transport. Even in 1928 only 10 million tons of freight travelled by water routes from the east coast to the west coast and in the reverse direction via the Panama Canal, whereas railways carried more than 250,000 million tons of freight. The speed of railway transport more than compensated for the lower cost per ton-mile of coastal shipping.

14. Foreign trade and American shipping.

(a) *Foreign shipping* was relied on to a great extent in transporting American exports and imports before the First World War. In 1801 90 per cent of American foreign trade was served by American shipping: in 1901 only 8·2 per cent of American exports and imports were carried in American bottoms. The relative decline was due to the following factors:

(*i*) The interior of the country had an attraction for investment.

(*ii*) A technological lead in shipbuilding was developed by Britain and Germany, in particular, which gave shipowners of those countries cost advantages.

(*iii*) German shipowners had the further advantage of subsidies which were denied to Americans until 1891, when a small mail subsidy was granted. British shipping lines benefited, too, from mail contracts.

(*iv*) The U.S.A. delayed conversion of her fleet from sail to steam.

(*b*) The First World War stimulated *American shipbuilding* and the growth of her shipping engaged in foreign trade: in 1920 43 per cent of American exports and imports were carried by her own ships and in 1935 36 per cent. By the end of the Second World War the U.S.A. was the leading shipping nation in the world, possessing about 60 per cent of the world's shipping.

ROAD TRANSPORT

15. The growth of the motor vehicle industry.

(*a*) *In 1914* the dominant form of transport in the U.S.A. was the *railroad*: the motor vehicle was of relatively minor significance. In 1895 four motor vehicles were produced in the U.S.A.: in 1921 1·5 million were produced and in 1929 4·8 million. Production diminished during the great depression to a low of 1·4 million in 1932, but in 1937 it again exceeded 4 million vehicles. Between 1918 and 1929 20 million vehicles were added to the nation's stock and another 20 million between 1936 and 1950. Approximately 44 million vehicles existed in the U.S.A. at the end of 1949 – 36 million cars and 8 million lorries and omnibuses.

(*b*) *After 1918* annual expenditure on motor vehicles exceeded expenditure on equipment for the railroads: the motor vehicle industry became as important an initiator of economic growth as the railroad had become in the nineteenth century. Demand for well-paved roads, petroleum, rubber, glass, iron and steel, leather, textiles, paint, polish, chemicals, service stations and hotels burgeoned.

(*c*) *Intervention* by the federal and state governments became necessary to provide adequate roads. State governments issued bonds and raised about $4 billion between 1920 and 1950. Taxes on motor vehicles and petroleum provided $2 billion a year by 1949, so that expenditure on roads grew to $800 million a year in the 1930s and $1·5 billion a year in the late 1940s.

16. The economic consequences of the use of the motor vehicle.

(*a*) *Short-distance passenger traffic* on railroads diminished: passengers were lost to private cars and to the omnibus companies. Carriage of freight by road increased so that by 1950 it was about one-sixth of that of railroad carriage in volume on inter-city routes.

(*b*) *Electric railways* gradually went out of service. (They can be compared with tramway services and trolley-bus services in Britain: these, too, were withdrawn as omnibuses became more reliable, economical and comfortable.)

(*c*) *Road traffic* grew at a rapid rate and, even in 1929, congestion on some city routes was a serious problem, baffling traffic engineers.

(*d*) A *more flexible transport service*, from door to door, encouraged the use of motor vehicles. Firms began to build up their own fleets of vehicles.

(*e*) The use of *hire-purchase* facilities expanded. People became more accustomed to the use of extended credit facilities. In the very short term this increased buying power, but in the long term it is possible that it diminished consumers' buying power because of the interest payments involved.

(*f*) The growth of the *oil industry* was stimulated.

(*g*) *Residential suburbs* of towns grew rapidly.

AIR TRANSPORT

17. Significance. Air transport became of greater economic significance in the second half of the 1930s when technology had advanced sufficiently to make air travel reasonably safe and speedy.

The number of passengers carried exceeded 3 million in 1940, 7 million in 1945 and 17 million in 1950 when the mileage for passenger and mail traffic exceeded 119 million, 240 million and 360 million miles (190 m., 384 m. and 576 m. km) respectively.

18. Government aid. Air-mail contracts helped subsidise American aviation and helped it grow into strong independent life: in 1945 more than 65 million (lb 29·5m. kg) mail was carried.

PROGRESS TEST 12

1. How important was water-borne transport in the U.S.A. before 1914? (**2, 3**)

2. How important were land grants in aiding railroad development? **(4)**

3. What impetus to economic growth did American railways provide? **(4–10)**

4. Discuss railroad consolidation. **(8)** What attempts were made to control railways by federal and state authorities? **(9, 11)**

5. Discuss the growth of shipping. **(12–14)**

6. How important was American shipping in the growth of overseas trade ? **(14)**

7. Assess the economic effects of the growth of the automobile industry. **(15, 16)**

PART SIX

LABOUR MOVEMENTS

CHAPTER XIII

LABOUR MOVEMENTS IN BRITAIN

1. Introduction: economic development and trade-union growth.

(*a*) *Conditions for growth.* Labour movements flourish best when the economic climate is favourable for their growth, that is when the rate of growth of demand for labour exceeds that of supply. Both the British and American economies have been transformed in the nineteenth and twentieth centuries by technological innovations, which have affected the demand for labour according to the nature of such innovations.

(*b*) *Sectoral differences.* Sectors of the economies have felt the impact of innovations at varying rates of progress so that the bargaining position of labour has tended to differ among the sectors at a given instant of time and over periods of time. Expanding industrialisation has been accompanied by a great increase in the demand for labour in the long period, but in some sectors innovations have, from time to time, caused a fall in the demand for labour.

(*c*) *Early industrialisation.* Initially, the transformation of both the British and the American economies from agrarian domination into economies dominated by industry, transport and trade involved vast investment programmes, making such demands on material and capital resources that the economic outlook for unskilled labour appeared bleak.

(*d*) *Maturity of growth.* In the long run, a growth of trade unions has been facilitated by technological innovations and the huge expansion of industry, accompanied by a growth in the size of business units.

(*e*) *Law and employers' attitudes.* Union growth has been fostered, too, by a liberalising of the law in relation to workers' combinations and by the increased willingness of employers to negotiate over pay, hours and conditions of work, both in Britain and in the U.S.A.

149

THE DEVELOPMENT OF LABOUR ORGANISATION IN BRITAIN BEFORE 1850

2. Obstacles to early growth. Most labour organisations in the early nineteenth century were small: combinations were local and sectional. Labour movements were restricted in their growth by the following factors:

(a) The *Combination Acts*.

(b) The small size of business units.

(c) A lack of willingness of employers to negotiate with unions.

(d) Poor transport and communication facilities making the organisation of labour beyond the confines of one locality somewhat difficult.

(e) A low standard of literacy among workers who tended to be ignorant of the benefits of collective bargaining on a wide scale.

3. Trade unions after 1824. The repeal of the *Combination Acts* in 1824 paved the way for a growth of trade unionism on a larger scale, despite the modifying Act of 1825 and the sharp economic recession which followed the boom of 1824–5.

(a) *The National Union of Cotton Spinners* was formed in 1829 by John Doherty, but it soon failed because the introduction of new spinning machines weakened the operatives' bargaining power.

(b) *The National Association for the Protection of Labour* was formed in 1830 and was a combination of more than a hundred trade societies, but this association crumbled in 1831 because of a lack of cohesion among its members, because of financial weakness and because of determined opposition by employers' organisations. Its main strength had been in Lancashire but Doherty's quarrel with the executive committee of the Manchester branch helped bring about the decay of this branch.

(c) *The Operative Builders' Union*, or General Trades Union as it was sometimes called, was built up by a federal organisation of local societies embracing seven types of craftsman in the building trades. One section of this union – the Operative Stone Masons – serves as a small example to illustrate the effect on union strength, as measured by total membership (not the sole criterion to use in assessing union strength), of fluctuations in the economy and the attitude of the state. (The estimates are those of J. H. Clapham.)

Date	Membership	Causes of changes
1833	6,000	Expanding economy.
1835	1,678	Campaign against union by the state.
1837	5,590	Trade conditions 1835–7 good.
1843	2,144	Depression 1838–42 and strike in 1841.
1852	6,700	Good trade conditions.

(In 1852 only approximately 10 per cent of stonemasons belonged to the union – probably because of the scattered nature and small size of building units in the nation.)

4. The Grand National Consolidated Trades Union.

(a) *Reasons for growth.* By 1833 the economy of Britain was on the upswing; there was a surge of expansion in the textile industries, the metallurgical industries, the building trades and transport development. An attempt was made to link workers' organisations in textiles, pottery, etc., into one large organisation of somewhat revolutionary intent – the Grand National Consolidated Trades Union, inspired partly, too, by disappointment with the electoral *Reform Act* of 1832.

(b) *Aims.* The union aimed to secure control of the means of production in the nation: a general strike was to be one of the weapons used in facilitating the carrying out of such a policy.

(c) *Difficulties.* The G.N.C.T.U. soon found itself in difficulties because:

(i) employers were antagonistic and combined to defeat the use of the strike weapon;

(ii) the laws of master and servant, conspiracy, etc., could be invoked easily against union members;

(iii) the state entered the arena against unions;

(iv) the union suffered from internal dissension;

(v) financial resources were inadequate.

(d) *Decline.* Towards the end of 1834 the G.N.C.T.U. began to wither away and its collapse marked the end of attempts to build up an all-embracing trades union aimed at social reform, if necessary by revolutionary methods.

5. Unions of the 1840s: background factors.

(a) By 1840 *small trade societies* were inadequate for the needs of labour in its effort to indulge in collective bargaining.

(b) *Industrial expansion* was being accompanied by an increase in the size of industrial units.

(c) *Expanding railway facilities* began to destroy local monopolies: markets tended to become regional rather than local. Collective bargaining needed to be organised beyond the local level. As labour mobility tended to increase so workers became more aware of economic trends outside particular localities.

6. The nature of unions in the 1840s.

(a) *Conservatism.* The unions of the 1840s tended to be less radical than those of the 1830s and confined themselves more to economic questions than their predecessors were wont to do. Whereas in the 1830s the great emphasis in the union movement was on the formation of an organisation embracing all trades, with social and political aims of great importance, the emphasis in the 1840s was on the creation of national unions for each trade.

(b) *Friendly society benefits.* Large unions appeared in mining, cotton spinning and engineering – mainly of a federal nature. These unions were less militant, had a sounder financial base and made use of permanent paid officials. Great stress was laid on friendly society benefits.

(c) *Origins.* Unions of this type originated largely after 1842 when the economy began to recover from the severe recession into which it descended after 1837–8. There was a brisk expansion of industry in 1834, 1844 and 1845 engendered by the railway boom.

(d) *1847 crisis.* The crisis of 1847 called a very brief halt to expansion and some unions had difficulty in surviving the last three years of the 1840s. Notably, the Miners' Association foundered as a national association, although mining unions survived in several counties.

THE "NEW MODEL" ERA

7. Social, economic and legal factors in union development. After 1850 there was a continued growth in the size of firms, helped to some extent by changes in company law with the passing of the *Limited Liability Acts* of 1856, 1858 and 1862. Large-scale employers had to be confronted by larger-scale trade unions if collective bargaining was to progress favourably. Skilled labour was relatively scarce; there were great improvements in communication and transport; after 1858 real wages of skilled workers, especially, rose steadily to reach in 1875 a level that was $33\frac{1}{3}$ per cent greater than in 1850; favourable changes in the law relating to trade unions were enacted; there was a significant increase in elementary education. Trade-union membership grew.

8. Changes in the law facilitating union growth. Important developments were as follows:

(a) The *Friendly Societies Act*, 1855, gave legal protection to the funds of societies with benefit functions.

(b) The *Molestation of Workmen Act*, 1859, specifically exempted peaceful picketing in trade disputes over wages and hours of work from penalties for molestation and obstruction.

(c) The *Trades Union Act*, 1871, (reversing the decision in *Hornby* v. *Close* (1867)), gave protection to trade-union funds against misappropriation.

(d) The *Conspiracy and Protection of Property Act*, 1875, legalised peaceful picketing again and reversed the *Criminal Law Amendment Act*, 1871, in this respect.

(e) The *Employers and Workmen Act*, 1875, gave employees equal legal status alongside employers.

9. Features of New Model unions. There arose after 1850 unions that were dubbed "New Models" by Sidney and Beatrice Webb. These unions had some salient characteristics in common:

(a) *They were national organisations* with a centralised form of control and administration.

(b) *Membership was limited* to legally apprenticed workers – the skilled workers.

(c) *Friendly society benefits* were offered and, in fact, were an important aspect of these unions.

(d) *Substantial subscriptions* were paid. This ensured that unions were financially stable and put membership beyond the reach of most unskilled workers.

(e) *Aims were almost wholly economic*, with settlements secured as far as possible by negotiation and arbitration.

10. Examples of New Model unions.

(a) *The Amalgamated Society of Engineers* (*A.S.E.*) was created out of 121 distinct societies in the engineering industry. Membership rose as follows:

1851.	12,000
1868.	33,000
1886.	52,000
1891.	71,000

(b) *The Boilermakers' Society*, which had 26,800 members by 1886, had Robert Knight, an industrial consultant, as secretary.

(c) *The Friendly Society of Ironfounders* concentrated on friendly society benefits.

(d) *The Amalgamated Society of Carpenters and Joiners* was ably led by Robert Applegarth, its general secretary, and had 23,000 members by the end of the 1870s. It was modelled on the A.S.E. (It should be noted that the roots of New Model unions can be traced back to the eighteenth-century trade clubs of skilled workers which amalgamated to form the highly centralised organisations found after 1850.)

11. The London Trades Council. In 1860 the unions formed the London Trades Council, which was dominated eventually by five leaders of the amalgamated societies. These were Applegarth (Amalgamated Society of Carpenters and Joiners), Allan (Amalgamated Society of Engineers), Guile (Friendly Society of Ironfounders), Coulson (Operative Bricklayers' Society), and Odger (Shoemakers' Society).

(a) *Efforts to effect co-ordination.* Through the London Trades Council attempts were made to co-ordinate the policies of unions with an emphasis upon the arbitration of disputes, thereby conserving the funds of the union movement.

(b) *The need for political action.* This was recognised, too, for some reform of the law was necessary to protect trade-union funds and to protect union members from legal action in the event of strikes taking place.

THE TRADES UNION CONGRESS

12. The birth of the Trades Union Congress.

(a) *Non-New Model unions.* By the early 1870s some 1 million workers were members of unions and the majority of these were not members of the amalgamated societies. The non-New Model unions were important in the boot-making, tailoring, printing, iron and steel and mining industries. These unions were important bodies in the development leading to the formation of the T.U.C.

(b) *Dualism.* For a brief period in the 1860s the London Trades Council and the amalgamated societies appeared to be moving away from the union movement in the north of Britain. The London Trades Council held aloof from the attempts to summon a national congress of trade unions in 1866, at which a body entitled the United Kingdom Alliance of Original Trades was set up, which, however, soon collapsed because of lack of financial support. The London Working Men's Association called a conference of trade unions in 1867 but the London Trades Council remained aloof and, in fact, organised its own Conference of Amalgamated Trades to advise the Royal Commission formed to enquire into the trade-union movement.

(c) *Protection of funds and reconciliation.* The London Trades Council was not represented at the Trades Union Congress of 1868, but Congress pledged itself to help the Council in efforts to secure legal protection for trade-union funds and reconciliation between the two groups followed.

(d) *Trade Union Bill.* In 1869 the London Trades Council sent delegates to the second annual conference of the T.U.C. Unity was established in time to enable the unions, collectively, to combat the government's Trade Union Bill, which gave legal protection to union funds, but made picketing and intimidation illegal.

(e) *The Parliamentary Committee.* In 1871 a Parliamentary Committee of the T.U.C. was formed and this became, eventually, a permanent institution. The Committee was formed to lobby M.P.'s for the amendment of the Trade Union Bill: it succeeded only in securing a division of the Bill into two separate ones.

(f) *Legislation and T.U.C. leadership.* Two Acts were passed in 1871 – the *Trade Union Act* and the *Criminal Law Amendment Act*. In the fight against the latter, leadership of the trade-union movement passed into the hands of the Parliamentary Committee.

(g) *Powers and success of T.U.C.* The T.U.C. was firmly established as the body through which unions would co-ordinate activities. However, it did not possess the powers that the American Federation of Labour possessed: it was merely a body through which unions could air their views and could discuss matters and it had no direct control over the actions of individual unions. By supporting the Conservatives at the election in 1874, the trade unions helped secure the repeal of the 1871 *Criminal Law Amendment Act*. The Parliamentary Committee also worked for amendment of the law of master and servant, which was secured in 1875.

NEW UNIONISM

13. Fundamental changes in union thought. The period 1875 to 1920 was one in which radical changes in the thinking of trade unionists emerged.

(a) *Friendly society activities.* These were now hived off from other union activities. Union contributions were regarded as contributions for union needs such as the support of strikes and not for welfare payments. Increasingly it was felt that the state should provide adequate welfare facilities.

(b) *Politics – the Labour Party.* Whereas unions before 1875 had largely been kept outside active politics, after the 1870s a close link was forged between the unions and the Independent Labour Party, later to become the Labour Party. Unions provided financial support and some of the Labour Party M.P.s were from the ranks of the unions.

(c) *Unfavourable legal decisions.* The need for intervention in politics was felt more keenly because of unfavourable judicial decisions against trade unions, because of unsympathetic employers and because of the attitude of successive governments:

(i) In the case of *M. Lyons v. Wilkins* (1896 and 1898) an injunction was granted to Lyons to prevent the Amalgamated Trade Society of Fancy Leather Workers picketing his premises. The Court of Appeal upheld the decision.

(ii) In 1901 the Taff Vale Judgment meant union funds were open to civil actions.

(iii) In 1901 the case of *Quinn* v. *Leatham* re-emphasised the Taff Vale decision.

(iv) In 1906 political actions safeguarded the union's position from above decisions. The *Trade Disputes Act* was passed.

(v) In 1909 in the Osborne Judgment it was declared *ultra vires* for unions to contribute to the Labour Party. This threatened to stifle the voice of the unions in Parliament because of lack of finance.

(vi) In 1913 the *Trade Union Act* legalised political contributions.

14. Union militancy. Unions tended to be more militant and more likely to resort to the use of the strike weapon. Notable strikes included the London Match Girls' Strike (1888) and the London Dockers' Strike (1889).

The number of working days lost through industrial disputes exceeded 17,000 in 1892, 30,000 in 1893 and 10,000 in 1898. From 1898 to 1908 the number of working days lost through industrial disputes was relatively small but the years 1908 to 1913 were marked by intense bitterness in the railway and coal-mining industries.

15. The general labour unions. The period after 1886 to the

First World War is often referred to as the period of New Unionism because of the rise of general labour unions catering for those workers outside the folds of the craft unions. These unions were organised on an industrial basis rather than on a craft basis. They had socialist and, often, revolutionary principles, but the socialist and revolutionary principles tended to thaw out after 1892: unionism was somewhat cautious in the first decade of the twentieth century.

Examples of general unions were as follows:

(a) Workers' Union (1898).
(b) Dock, Wharf, Riverside and General Labour Union (1889).
(c) Tyneside and General Labour Union (1889).
(d) Amalgamated Society of Gasworkers and Brickmakers (1889).
(e) National Amalgamated Union of Labourers (1889).
(f) Gas Workers' and General Labourers' Union (1899).
(g) National Federation of Women Workers (1906).

16. The craft unions.

(a) *Craft unions* continued to be of great importance after 1889 in the engineering, building and shipbuilding industries. In 1892, of an estimated 1,570,000 trade-union members, about 1 million were, in fact, skilled workers.

(b) *General unions* did not have the attraction sometimes supposed and they were not "new": they had predecessors in the 1830s, and in the third quarter of the nineteenth century unions outside the amalgamated societies were of considerable importance.

17. Arbitration and conciliation boards. No compulsory arbitration of industrial disputes existed in the nineteenth century, but schemes initiated by entrepreneurs such as Mundella were forerunners of conciliation and arbitration boards.

The *Conciliation Act* was passed in 1894: it empowered the Board of Trade to form boards of arbitrators and conciliators on the request of the parties in dispute. By 1913, 325 arbitration and conciliation boards were in existence.

TRADE UNIONS IN THE EARLY TWENTIETH CENTURY

18. The Trade Disputes Act, 1906. The Labour Representation Committee, reformed as the Labour Party after twenty-nine

M.P.s had been returned at the General Election of 1906, formed the Parliamentary Labour Party, which pressed the Liberals for reform of trade-union law.

(a) *Main provisions of the Act.* The *Trade Disputes Act* safeguarded unions:

(i) Trade unions were protected from lawsuits where none could have arisen from actions performed by an individual.

(ii) Peaceful picketing was legalised, as were non-violent actions in restraint of trade.

(b) *Militancy.* The Act ushered in a period of union militancy. From 1902 to 1908 the cost of living rose by approximately 5 per cent and from 1908 to 1913 by about 9 per cent. Pressure was brought to bear on union leaders to secure wage increases. Marxism took a greater grip on workers in industry and the ideas of syndicalism fell upon more fertile ground, particularly in South Wales.

(c) *National strikes.* In 1911 a national strike on the railways occurred, to be followed in 1912 by a national strike in the coal industry and an attempt by the Transport Workers' Federation to bring about a national strike of transport workers, occasioned by a dispute with the Port of London Authority. Ideas of a general national strike took firmer root and attempts were made to build an alliance of miners, railwaymen and general transport workers to forge a strike weapon of great menace to put pressure on the government, but the onset of the First World War delayed the use of such a weapon.

(e) *Employer–union antagonism.* Faced with the rise of more and more effective foreign competition in both the domestic market and in markets abroad, employers sought to reduce costs of production more effectively and reacted antagonistically to demands for wage increases, particularly in coal-mining and in transport. The seeds of industrial disputes of the 1920s were sown in the years between 1906 and 1914.

UNIONS AND THE FIRST WORLD WAR

19. The effects of the First World War on unions. The period 1914–18 was one of relative industrial harmony because of a more effective dialogue between the government and the unions, and because of legislation giving the state greater control over labour. However, some divisions appeared in the ranks of labour:

the rank and file became suspicious of a leadership closely allied with the government. Shop stewards acquired a greater dominance over union members at local level: the shop stewards' movement was at the centre of strikes that were called in the engineering and munitions industries in 1917.

(a) The *Munitions Act*, 1915, introduced compulsory arbitration: in 1917 the new *Munitions Act* made wage awards national ones. The Committee on Production was the abitration tribunal in the munitions industry: its awards were applied to other industries – shipbuilding, chemicals, toolmaking, etc.

(b) Under the *Wages (Temporary Regulation) Act*, 1918, the Committee on Production became the Court for Arbitration and, in the following year, it became the Industrial Court. Out of the First World War, therefore, emerged machinery for settling disputes by arbitration and for changing wages nationally.

THE INTER-WAR PERIOD

20. The Restoration of Pre-war Practices Act, 1919.

(a) *End of government arbitration.* The relinquishing by unions of trade practices under the Treasury Agreement and the *Munitions Act* ended. There was a return to bargaining without state enforcement of decisions in industry (although railway and coalmines remained under direct government control).

(b) *Minimum wages.* A marked difference from pre-war conditions existed. Minimum wages were enforceable in a much wider range of occupations.

21. The Whitley Councils. Whitley Councils were instituted as a result of the Report of the Whitley Committee, but only in spheres of government employment was a successful construction of such negotiating machinery carried out.

22. The post-war boom. Trade-union membership had increased to some 8·3 million by 1920, helped partly by union successes during the war and by the great boom in the economy after 1918. For a brief period after the war there was a release of pent-up demand for goods in the domestic market and in the export markets. Full employment existed: returning soldiers were quickly absorbed in the labour force. Halcyon days appeared at hand for the British economy. Times, therefore, were propitious for trade unions to increase their membership: 1919 was a year of

considerable militancy among workers in the mining and railway industries, and there was trouble even among members of the police forces, strikes occurring in London and Liverpool. Nearly 35 million working days were lost through industrial disputes in 1919.

23. The General Strike and years of high unemployment.

(a) *Increasing militancy was curbed* in 1920 by lack of union cohesion, firm government action and a severe recession in the economy. By 1922 more than 1·5 million were registered as unemployed.

(b) *Industrial relations in coal-mining were embittered* by the return of the coal-mines to private enterprise, in the face of the Sankey Report, and by colliery owners' attempts to cut wages. Unsatisfactory relationships in the coal-mining industry were exacerbated by increasing difficulties in export markets caused by:

(i) a restoration of the pound to the gold standard at a parity estimated to be 10 per cent too high;
(ii) relatively high costs in the British coal industry;
(iii) the opening of new and more efficient mines in Europe;
(iv) increasing economies in the use of coal as coal-burning appliances of greater efficiency were developed.

(c) *A Royal Commission* appointed to enquire into the difficulties in the coal-mining industry recommended a wage reduction, which was rejected by the miners, who were locked out on 26th April 1926.

(d) *A general strike was called*, but this was a failure: it was called off after unsatisfactory negotiations.

24. Results of the General Strike. The government was alarmed at the danger to the economy that arose from a general strike, and a curbing of trade unions' power was deemed necessary.

(a) *Trade Disputes and Trade Unions Act*, 1927. The provisions of the Act were as follows:

(i) Sympathetic strikes were declared illegal.
(ii) The intimidation of workers became a criminal offence.
(iii) Civil servants were prohibited from joining unions affiliated to the T.U.C., but existing membership could be retained.
(iv) The political levy could be paid only by "contracting in."

(b) *Moderation*. Trade-union leaders became more moderate in the face of strong government attitudes and continuing large-scale unemployment. The general strike, regarded as the ultimate weapon, was viewed with scepticism as an instrument of labour's policy: sympathy with syndicalism declined. Trade-union leaders tended to follow somewhat conservative policies in the 1930s because they wished to avoid being considered too radical in attitude.

(c) *Fall in membership.* Membership declined to a nadir of 4·4 million in 1932.

25. The resurgence of unions.

(a) *Unemployment* was considerable in the 1930s but it was confined largely to coal-mining, textiles, iron and steel and ship-building. The actual number in employment continued to increase, especially in the Midlands and the South-east, where motor vehicle, electrical goods, radio and allied industries expanded. Union membership increased to 6·3 million by 1939.

(b) *The rise of leaders* such as Bevin and Citrine and the sympathetic attitude of employers such as Sir Alfred Mond advanced the status of the T.U.C., which was drawn into consultation on the nationalisation of the textile industry and the fishing industry. The T.U.C. was consulted, too, on plans for mobilisation and for air-raid precautions.

TRADE UNIONS, 1939–50

26. State intervention in the Second World War.

(a) *Control over labour.* Close government control over labour was exerted under:

(i) the *Emergency Powers (Defence) Act;*
(ii) the Restriction on Engagement Order;
(iii) the Essential Work Order.

(b) *Control over employers.* Employers, too, were controlled by the state in their engagement and dismissal of labour, but collective bargaining over wages and conditions of work continued.

(c) *Responsibility of unions.* Trade-union leaders were given positions of great responsibility, and the involvement of Russia in the war probably killed any possibility of the development of a rank-and-file movement like that which occurred in 1914–18.

(d) *Rise in membership.* There was a tremendous increase in trade-union membership to 8 million by 1945.

27. The post-war years.

(a) *Rise in membership.* The growth in membership continued. By 1955 some 10 million workers belonged to trade unions. At the same time the number of unions declined as a more effective federation of unions developed.

(b) *Repeal of 1927 Act.* In 1945 a Labour government was elected and the 1927 *Trade Disputes and Trade Unions Act* was repealed.

(c) *Nationalisation.* The nationalisation of the transport, gas, electricity and coal industries and the Bank of England gave the state greater control over the economy than it had possessed hitherto. The trade-union movement lost many of its leaders to the public corporations that were instituted, and found the immediate post-war years somewhat difficult. Obviously, the Labour government was supported by the union movement, but at the same time unions had to remain faithful to the system of collective bargaining for their members, despite the T.U.C.'s agreeing to co-operate in a policy of wage restraint. In 1950 voluntary wage restraint, in the face of a rising cost of living, was abandoned. Trade unions could no longer "hunt with the hounds and run with the hare." In 1968-9 they faced a similar dilemma.

PROGRESS TEST 13

1. Outline the growth of the New Model unions. (**7–10**)

2. Show how the T.U.C. originated. (**12**)

3. Why was there unrest among trade unions and employers between 1890 and 1914? (**13–17**)

4. Assess difficulties in industrial relations in the inter-war period. (**20–25**)

5. Discuss the causes and effects of the General Strike of 1926. (**23, 24**)

6. How did the Second World War affect trade unions? (**26–27**)

CHAPTER XIV

LABOUR MOVEMENTS IN THE U.S.A.

THE NATURE OF TRADE-UNION DEVELOPMENT, 1850–1950

THE history of labour movements in the U.S.A. falls conveniently into four periods of study in the years 1850–1950.

1. The years before the outbreak of the Civil War. In the first half of the nineteenth century trade unions were in an embryonic stage of development. Although industrialisation grew more rapidly after 1820, the economy, even in 1860, was still predominantly agrarian: business units were relatively small. Consequently trade unions, too, tended to be relatively small and essentially local in character, although a few unions aspired to national status as improved communications facilitated union growth.

2. Labour movements, 1860–86.

(*a*) *The Civil War.* There was a considerable growth in trade unions during the Civil War, which also brought along inflationary pressures, caused supply bottlenecks to form and left some demands unsatisfied. Workers were caught up in the spiral of prices: in the Aldick Report it was calculated that, whilst money wages increased by 43 per cent between 1860 and 1865, prices increased by 117 per cent; real wages fell by some 34 per cent. As a result of this, workers organised on a larger scale and there was a marked increase in the number of national trade unions; about thirty-two unions were organised on a national basis by 1865.

(*b*) *The period after* 1865. Political reform attracted labour organisations after the Civil War until the depression of 1873–8 thrust trade unions into a trend of dwindling membership and dwindling financial resources. The lifting of the depression in 1879 encouraged union growth once more: efforts were made to promote larger national labour organisations, culminating in the formation of the American Federation of Labour in 1881, and the rapid growth of the Knights of Labour.

3. The period of A.F.L. emergence and dominance, 1886–1935.

(a) *Influence of Samuel Gompers.* Although there was a great interest displayed by the A.F.L. in politics in its very early stages of development, under Samuel Gompers policies not unlike those of the New Model unions of Britain in the earlier years were adopted. By 1900, despite the 1893–6 depression, the use of injunctions and private detectives by employers, and defeat in the iron and steel and railway industries, the A.F.L. had negotiated with many employers and secured acceptance of its collective bargaining policies over a wide range of industries. Membership exceeded 1 million by 1901, 2 million by 1915 and soared to more than 4 million by 1920.

(b) *Decline in membership.* In the economic boom of the period 1922–9, membership declined, partly because the A.F.L. failed to adapt itself sufficiently rapidly to changing economic and industrial conditions and partly because there was a growth of company unionism. There was a further fall in membership during the great depression of 1929–38.

4. The Congress of Industrial Organisation and the American Federation of Labour. After 1935 a system of dual unionism existed:

(a) *Union growth.* Under the shadows of the *Norris–La Guardia Act*, the *National Industrial Recovery Act* and the *National Labour Relations Act*, union membership increased rapidly from about 3 million in 1932 to 9 million in 1940.

(b) *Dualism.* The A.F.L. gave birth to the Committee for Industrial Organisation in 1934, which became a separate body (named the Congress of Industrial Organisation in 1938) aiming to unionise labour on an industrial rather than a craft basis. By 1937 the C.I.O. had more members than the A.F.L.

(c) *Union membership.* By 1945 both the C.I.O. and A.F.L. each had a membership of more than 6 million and had become powerful institutions in American society.

(d) *Affiliation.* In the 1950s the A.F.L. and C.I.O. drew closer together in a loose affiliation. Huge business corporations had to face a huge organised labour complex with unions aspiring to management status in industry.

EARLY TRADE-UNION DEVELOPMENT, 1830–60

5. City unions. Initiated by skilled workers of Philadelphia, who formed the Philadelphia Mechanics' Union of Trade Associations in 1928, craft unions developed between 1828 and 1832 in the main cities. They included workers of more than one craft organised on a city-wide basis and were reformist in character, their demands ranging from a reduction in the hours of work, a restriction of child labour and payment in a stable currency, to free school facilities, direct election of public officials, etc. One legal anomaly campaigned against was the liability of people to be imprisoned for debt (making poverty a crime) and the lack of legal redress against employers who failed to pay wages to the workers. Unions made use of the purchasing power theory in supporting demands for higher wages.

6. Changes in union policies and union growth. Trade societies turned again to economic rather than political action. After 1832 conditions in the middle years of the 1830s favoured union growth:

(a) There was an *increase in prosperity* accompanied by a rising demand for labour, especially skilled labour.

(b) *Transport facilities* improved, making a more widespread organisation of labour more feasible.

(c) *The cost of living* increased more rapidly than wages, making combined action imperative if workers were to maintain real wages at a reasonable level.

(d) *Workers' solidarity* grew in the face of employers' using more female labour and immigrant labour.

7. Efforts to promote a national union.

(a) *National Trades Union.* In 1834 local societies from cities such as New York, Boston, Philadelphia, etc., met to establish the National Trades Union. This union was unlike the Grand National Consolidated Trades Union formed in Britain. The latter aimed to take over the means of production by economic and political action, whilst the former was interested only in economic objectives to be pursued by economic action. To the alarm of the business community frequent use was made of the strike weapon: between 1833 and 1837 some 168 strikes occurred in New England alone.

(b) *Setbacks.* The union movement suffered a setback in 1831 when the cordwainers were successfully prosecuted for combining to raise wages: such a combination was held to be "a conspiracy against the spirit of the common law." Unionism was dealt a severe blow by the acute economic depression of 1837: the National Trades Union collapsed.

8. Reformist activities. Following the collapse of trade unions in the economic depression of 1837, reforms of various types attracted support from labour. The most notable movements were those of Charles Fourier (with his phalanxes and phalansteries) and the Agrarian League of George Harry Evans.

9. Commonwealth v. Hunt (1842). In 1842 the right of workers to join unions was recognised. Chief Justice Lemuel Shaw declared that neither combinations of workers nor their striking for a closed shop were in themselves unlawful conspiracies. Employers, however, continued to attack unions in the courts, involving unions in costly litigation.

10. Revival of unions. Prosperity increased in the 1850s as transport improved, facilitating an expansion of trade (the U.S.A. had more than 30,000 miles (48,000 km) of railway track laid by 1860). At the same time the whole of trade received additional monetary lubrication from an increase in the world's supply of gold arising from gold discoveries in California and Australia. Local and national unions grew with renewed vigour. Some ten national unions were formed; four examples of these may be cited:

    (a) The Typographical Union (1850).
    (b) The National Finishers' Union (1854).
    (c) The National Union of Machinists and Blacksmiths (1859).
    (d) The National Moulders' Union (1859).

11. Salient features of unions of the 1850s.

    (a) They were largely *craft unions* confined to skilled workers: the organising of immigrants and of unskilled workers proved difficult.

    (b) *Sickness benefits* were of importance as in the New Model unions in Britain.

    (c) *Strike funds* were carefully built up and the strike weapon was used when deemed necessary. The *New York Tribune* in April 1854 observed that strikes were frequent in the 1850s and that public opinion was often in support of strikers. In 1860 an extensive strike occurred in New England involving workers in twenty-five towns in support of demands for higher wages, to keep labour's rewards in step with the rising cost of living.

(d) The *national unions* that developed were unions confined to particular trades: each union consisted of workers from one trade: the unions were trade unions and not trades unions.

(e) Wherever possible disputes with employers were settled by *collective bargaining*. The unions were realistic business-like organisations.

TRADE UNIONS, 1860–73

12. The Civil War. The war had a disruptive effect on the economic life of the nation. Real wages fell considerably, the professional-class employees, government employees and women workers being the worst affected. Wages were kept down by employers by the use of labour-saving machinery, by the employment of immigrants (some 800,000 entered the U.S.A. during the Civil War) and by increased use of the labour of women and children. Lower-paid workers were adversely affected by the government's use of indirect taxation to raise revenue. Strikes for increased pay broke out in 1863 and there was a surge forward in the organising of labour.

13. The emergence of national unions.

(a) *Factors encouraging national unions.* There was a marked increase in the number of national unions stimulated by economic development:

(i) Competition increased from *immigrant labour* brought in under the *Contract Labour Law* of 1864, which permitted an advance of money to prospective immigrants in return for a lien upon their wages. Native labour had to organise on a large scale to combat this threat.

(ii) *Industrial changes* resulted in increased division of labour with a greater use of semi-skilled and unskilled labour. Local skilled labour found itself threatened by migratory unskilled labour.

(iii) *Markets* became more national in character. Wages scales began to be linked on a national basis, necessitating unions organising on a national basis.

(iv) *Other factors* such as improved transport and communications, better educational facilities and increased size of business units facilitated larger-scale labour movements. By 1873 economic forces were bringing about the emergence of very large-scale firms. Labour could bargain successfully with large industrial and commercial complexes only if it, too, could speak from a position of strength.

(b) *Examples of national unions*. National labour unions arose in the construction industry, among railroad workers, cigar workers, iron and steel workers, shoemakers, miners, etc. By 1871 there were more than 300,000 workers in national labour unions but it must be emphasised that this was only a small proportion of the total labour force. Among the thirty-two national labour unions that arose in this period three of note must be mentioned:

(i) *The National Labour Union*. This was essentially a political reform organisation which was inaugurated in 1866 at Baltimore. Its main aims were: (1) to obtain an eight-hour working day; (2) by legislative reform, to obtain greater control for workers over industry; (3) to set up producer co-operatives; (4) to settle disputes wherever possible by negotiation, arbitration and conciliation, using the strike weapon only as a last resort; (5) to obtain the establishment of a government bureau of labour.

(ii) *The Knights of St Crispin*. This union was founded at Milwaukee in 1867 to provide a large-scale labour organisation among shoemakers. Like the National Labour Union, it foundered during the panic of 1873, and was absorbed by the Knights of Labour.

(iii) *The Knights of Labour*. Unak S. Stephen helped form this labour organisation. Its aims were modern in aspect – an eight-hour day, income and inheritance taxes, provision of compensation for those injured at work, the setting up of a post office savings bank, the instituting of income tax and inheritance tax, etc. The union's main growth came after 1873 and is referred to in **15** (a) below.

14. Economic crisis, 1873: union failures. In September 1873 the financial and business empire of Jay Cooke (then the U.S.A.'s most noted financier) collapsed, bringing in its train a closure of banks and a temporary halting of the activities of the stock exchange. Some sectors of the economy bumped along the floor of a recession for some six years after Jay Cooke's failure. Unions, not unexpectedly, found survival difficult: falling prices, falling production, business stagnation and considerable unemployment do not promote union growth. By 1877 only nine national unions were still in existence and membership had fallen to 50,000. Unions lacked the basic structural strength to ride out recessions of this nature; they lacked both capital and managerial "know-how."

1873–86: THE A.F.L. AND THE KNIGHTS OF LABOUR

15. The emergence of the American Federation of Labour. Two important organisations survived the economic difficulties of 1873 – the *Knights of Labour* and the *Cigar Makers' Union* (from which the American Federation of Labour was to emerge).

(a) *The Knights of Labour.* Since the organisation operated in some secrecy until 1879, and employers took drastic action against unionists involved in disputes, the Knights of Labour attracted defectors from other unions in the 1870s. Until 1884, the union avoided as far as possible the use of strikes and boycotts, but in 1885 a successful strike was called on the Gould railway system and membership of the organisation grew to 700,000. In 1886 the union was involved in another strike against the Gould organisation and was compelled to surrender. Public opinion became increasingly hostile towards the Knights of Labour following the Chicago Haymarket bomb outrage. Its hegemony over the American labour movement began to wither away because:

(*i*) it contained within itself too many conflicting elements;

(*ii*) social reform rather than trade-union organisation attracted too much of its attention;

(*iii*) unsound financial policies were followed – no adequate strike fund was built up;

(*iv*) in the public mind it was associated with violence.

(b) *Formation of the American Federation of Labour.* In 1881 the Federation of Organised Trades and Labour Union was formed, the Cigar Makers' Union and the Carpenters' Union being prime movers in this formation. In 1886 the union was reformed as the A.F.L., with Samuel Gompers as president. Dissatisfaction with the Knights of Labour's social reform policy was a cause of the founding of the A.F.L. in which business unionism was much more to the fore. Membership was limited to craft and labour unions. In 1886 the A.F.L. had about 138,000 members but it was to grow rapidly during the next thirty-four years.

16. Employers' attitudes to unions: blacklists. American entrepreneurs were less inclined to accept collective bargaining than their British counterparts, preferring to make contracts with individual employees and to insist on "yellow-dog" contracts in which employees agreed to refrain from becoming members of unions. Blacklisting of union leaders was a popular weapon. The lock-out was frequently used against strikes. Use was made of spies (often private detectives) to infiltrate unions. Pinkerton's Detective Agency played a prominent part in activities of employers to frustrate American labour movements.

TRADE UNIONS, 1886–1935

17. The American Federation of Labour, 1886–1914.

(a) *Acceptance of capitalism.* By 1897 the A.F.L. had established itself as leader of the American labour movement. It gained considerably in membership from the failure of the Knights of Labour; membership trebled between 1900 and 1908 and by 1914 it had 2 million members. Its leaders were generally non-political, despite socialist pressures from within. Gompers successfully kept the A.F.L. on the non-political line in the nineteenth century, his main setback being in 1894 when he temporarily lost the presidency. Gompers' view was that labour lived under the wage system in a capitalist society and that it was the job of the A.F.L. to secure as big a slice of the national cake as possible for its members. In the first decade of the twentieth century, however, the A.F.L. became more active politically: in 1906 it presented a Bill of Grievance to the President, the Senate and the Speaker of the House and in 1908 presented a list of demands for reform to both parties. The A.F.L. members worked openly in support of Bryan at the 1908 election and supported Wilson four years later.

(b) *Female unionists.* A notable feature of the union movement is the interest displayed by female workers in the labour movement in the twentieth century. The International Ladies' Garment Workers' Union and the Amalgamated Clothing Workers' Union contained not only large numbers of women but also large numbers of immigrant workers. Industrial unionism was a feature of these unions, even within the ambit of the A.F.L., and persisted, too, in mining and among dockyard workers.

18. The A.F.L. in the First World War.
Relatively, labour became a scarcer factor of production. Demand from Europe for American products caused an expansion of production, just as the flow of immigrants declined and many workers left the U.S.A. to return to Europe. The position was exacerbated when the U.S.A.

entered the war, some 4 million men joining the armed forces.
The A.F.L. supported the war effort and, just as British trade-union
leaders occupied prominent positions in official spheres, so labour
leaders in the U.S.A. were offered positions on war boards.
Gompers was placed on the Advisory Committee of National
Defence. In 1918 a War Labour Board and War Labour Policies
Board was set up: the former was to adjudicate between labour
and employers. A Railway Wage Commission looked after wage
adjustments on the railroad. Such bodies found collective bargaining facilitated by the existence of unions and tended to favour
union development. Membership of the A.F.L. increased to more
than 4 million and included professional workers.

19. Reconstruction. The American Federation of Labour had a
detailed programme of reconstruction in which it made demands
for labour in 1919. The demands that it wanted implemented
were as follows:

(*a*) A two-year cessation of immigration.
(*b*) Full employment policies.
(*c*) A guarantee that all employers would possess the right
to bargain collectively.
(*d*) Government ownership of some public utilities.
(*e*) Legislation to curb the power of corporations.
(*f*) The setting up of government employment agencies.
(*g*) Assistance for workers to purchase their own homes.
(*h*) The abolition of child labour.
(*i*) The establishment of equal pay for women and men.

20. Obstacles to unionism. The A.F.L. had to contend with
difficult circumstances that prevented the attainment of the above
objectives:

(*a*) The courts were unsympathetic.
(*b*) Some 4 million members of the armed forces were thrown
on the labour market soon after the war ended, weakening
labour's bargaining powers.
(*c*) There was a sharp recession in the economy during 1920–1.
(*d*) Employers' associations displayed a hardened attitude to
unions, the *agent provocateur* was used, as was propaganda
against labour movements, and great play was made on the fears
of communism, which created a national suspicion of radical
tendencies.

(*e*) Rank-and-file loyalty to unions was loosened by the use of company welfare schemes, profit-sharing ideas, the sale of stocks and shares to employees, and the setting up of shop committees.

(*f*) The A.F.L. failed to adapt itself to changing conditions. It adhered too rigidly to craft unionism instead of changing its policy to one of industrial unionism to provide workers in the "new" mass-production industries.

21. Decline. Despite a period of economic boom from 1921 to 1929 in most sectors of the economy, which kept unemployment at a relatively low level, in 1929 the membership of the A.F.L. was less than it had been in 1920. This trend was contrary to that which would be expected normally – partly, because unions became too interested in speculation and capital gains rather than being primarily concerned with building up union membership.

22. The 1929 crash and the onset of the New Deal.

(*a*) *Membership.* Since the A.F.L. failed to expand in times of rising economic prosperity it was hardly to be expected that membership would increase during the great depression. In fact membership did not fall off to the extent that events might have warranted: even in 1932 total union membership was almost 3 million despite unemployment figures of 16 per cent in 1931, 24 per cent in 1932 and 25 per cent in 1933.

(*b*) *Strengthened legal position.* Legislation to facilitate the organising of labour for collective bargaining by the enacting of the *Anti-Injunction Act*, 1932, the *National Industrial Recovery Act*, 1933, and the *National Labour Relations Act*, 1935, placed unions in a stronger legal position than hitherto, providing them with a base from which to expand.

THE INDUSTRIAL WORKERS OF THE WORLD

23. Syndicalism in the American labour movement.

(*a*) *Aims of the I.W.W.* The I.W.W. was inaugurated in 1905. It rejected the idea that workers could progress within the capitalist system: its view was that labour's only hope was to secure control of the means of production. A radical programme of union organisation was drawn up: local unions were to be joined into national unions, with thirteen departments to co-ordinate the activities of national unions in related industries.

(*b*) *Division among the leaders.* A split occurred which weakened the I.W.W. so that it did not become a real threat to the A.F.L. In 1908 two groups emerged:

(*i*) De Leon's revolutionary faction.

(*ii*) A non-political group which favoured direct action and became the core of the I.W.W. This group took part in notable strikes, *e.g.* (1) the Pressed Steel Car Company Strike, 1909 (successful); (2) the Textile Workers' Strike at Lawrence, Massachusetts, 1912 (successful); (3) the Paterson Strike, 1913, which lasted twenty-two weeks and depleted the I.W.W.'s funds.

24. Opposition to war. The I.W.W. virtually died when it opposed American entry into the First World War: it was attacked by employers and the A.F.L. Gompers saw the I.W.W. as a potentially dangerous rival that had to be crushed. Nevertheless, the I.W.W. did provide aggressive trade-union leadership in the spheres where labour needed organising. The "Wobblies" – unskilled migratory workers – and negroes were helped by the industrial unionism of the I.W.W. At the height of its power the I.W.W. probably had no more than 60,000 members at a time when the A.F.L. had more than 2 million.

THE LAW AND LABOUR MOVEMENTS

25. The use of injunctions.

(*a*) *Frustration of unions.* In the 1870s, judges issued injunctions to frustrate union activity in the case of railroads in receivership, and in the 1880s began to issue injunctions for railroads not under the jurisdiction of courts.

(*b*) *Prohibition of activities.* In the Chicago Railway Strike of 1894, an injunction was obtained prohibiting a wide range of activities by the American Railway Union. The injunction was based on the *Sherman Anti-Trust Act*, 1896, which declared illegal conspiracies to restrain trade and commerce.

(*c*) *Prohibition of boycott.* In 1906 the president of the Buck's Stove & Range Co. of St Louis obtained a federal injunction to stop the A.F.L. blacklisting or boycotting the firm's goods, following strife between the company and employees. Gompers was sentenced to gaol for contempt of court, but never served his sentence.

26. The Sherman Anti-Trust Act. The *Interstate Commerce Act*, 1887, and the *Sherman Anti-Trust Act*, 1890, gave the courts something more tangible than the common law on which to base decisions when requests for injunctions were considered. The latter Act was more important. It contained vague provisions that could be used against trade unions:

(a) The *Sherman Anti-Trust Act* declared illegal contracts, combinations or conspiracies *in restraint of trade* among "the several states or with foreign nations" (Section 1).

(b) The *circuit courts* of the U.S.A. were empowered to prevent and restrain violation of the *Sherman Act*. Pending the outcome of cases the courts could make restraining orders, or issue prohibitions as circumstances dictated.

(c) Persons whose business interests were injured by action in violation of the Act had the right to *sue for damages*.

27. Adverse court decisions for the labour movement.

(a) *Boycott damages.* In 1902 the *United Hatters of North America* had to pay substantial damages to Loewe & Co. of Danbury for boycotting the company's products.

(b) *The Buck's Stove & Range Co. case* (1906). (*See* **25**(c).)

(c) "*Yellow dog*" *contracts.* In 1917 unions were denied the power to intervene in "yellow dog" contracts between employer and employees.

(d) *Strike damages.* In 1919 the United Mineworkers had to pay damages to the Coronado Coal Co. for calling a strike.

(e) *Closed shop declared illegal.* In 1921 the action of the International Association of Machinists in trying to force a closed-shop policy upon the Duplex Printing Press Co. was declared illegal.

(f) *Picket limitation.* In 1921 the Supreme Court ruled that picketing could be limited to peaceful picketing and that courts could limit picketing to a single picket.

28. The Clayton Anti-Trust Act, 1914. This Act was regarded as a "charter of liberties" by organised labour.

(a) *Section 6* declared that nothing contained in the anti-trust laws should be so construed as to forbid the existence of labour organisations.

(b) *Section 20* stated that no restraining order or injunction should be granted by the courts in cases arising from disputes over conditions and terms of employment except where property rights could not otherwise be protected.

However, the Act failed to prevent the use of injunctions against organised labour.

29. The Norris–La Guardia Act, 1932.

(*a*) "*Yellow dog*" *contracts* were declared to be unenforceable in the courts. (Section 3.)

(*b*) It was declared that in cases involving labour disputes no court was to have the power to issue *restraining orders* or *injunctions* to stop workers belonging to unions, to stop workers striking, to stop workers assembling peacefully, etc. (Section 4.)

(*c*) *The power of the courts* to issue restraining orders or prohibitions on the grounds that workers were engaged in unlawful combination or conspiracy was removed. (Section 5.)

(*d*) *Individual members* of unions were declared to be not responsible for the actions of officers of unions or for other members. (Section 6.)

(*e*) *Injunctions* could be granted only where it appeared very probable that unlawful acts would take place, or where substantial and irreparable injury to a plaintiff would follow:

(*i*) if the denial of an injunction appeared likely to inflict greater harm upon the complainant than its granting would do upon the defendant;

(*ii*) if the complainant had no adequate remedy at law; or

(*iii*) if the public officers charged with the duty of protecting the complainant's property were unable to do so.
(Section 7.)

LABOUR UNDER THE NEW DEAL AND AFTER THE SECOND WORLD WAR

In the 1930s labour movements at long last gained full legal recognition. They attained the legal status that British trade unions achieved in the years 1906 to 1913.

30. The National Industrial Recovery Act, 1933. Section 7 of the Act stated the following:

(*a*) Employees should have the right to organise and bargain collectively, free from interference, restraint or coercion.

(*b*) No employee as a condition of employment should have to join a company union or refrain from joining a labour organisation of his own choice.

(*c*) Employers would comply with the maximum-hours requirement, minimum rates of pay and other conditions of employment approved by the President.

31. The National Labour Relations Act, 1935. This Act clarified and safeguarded further still the position of organised labour. It noted that the refusal of employers to recognise organised labour led to industrial strife, inefficiency and low wages. In this way the flow of trade was impaired. The Act stated that the policy of the American government, henceforth, was to encourage labour to organise collectively and to indulge in collective bargaining.

(*a*) *Employers and unions.* The Act declared as unfair certain labour practices by employers:

   (*i*) Interference in labour's guaranteed rights.
   (*ii*) Attempts to dominate the formation or administration of a labour organisation or to contribute to it financially.
   (*iii*) To refuse to bargain collectively with employees' representatives.
   (*iv*) To discriminate against an employee who gave testimony under the Act.
   (*v*) To discriminate in favour of or against a particular labour organisation.

(*b*) *The National Labour Relations Board.* This board was empowered to prevent any person engaging in unfair labour practices. The power granted was exclusive, but there was to be right of appeal to circuit courts of appeal.

32. The Social Security Act, 1935.

(*a*) *Welfare provisions.* The Act included welfare provisions for old age, unemployment and dependent children.
(*b*) *The Social Security Board.* This board was established to raise revenue by means of taxes on employers and to supervise the payment of grants under the Act to the various states, which had to have stated provisions in their laws before grants could be paid.

33. The Taft–Hartley Act, 1947. The militancy of the C.I.O., in particular, and the A.F.L., gave rise to concern about the state of the law regarding labour organisations. It was considered that the 1935 *Labour Relations Act* gave unions too much freedom. The continued growth of unions from 1934 made them into large, complex organisations that could paralyse the economic life of the nation if strikes occurred in key sectors of the economy. In 1947 the *Taft–Hartley Act* (or *Labour–Management Relations Act*) decreed that the President had power to order strikers to cease striking for a period of eighty days, if the strike was in a key sector of the economy. Attempts were to be made during the eighty days to achieve a settlement. Two other important provisions were as follows:

(*a*) *No closed shop*. Individuals were given the right to refuse to join unions, which were not allowed to insist on a "closed shop."

(*b*) *Unfair labour practices*. The Act laid down what was referred to as unfair labour practices, which were declared unlawful and restrictions were imposed on contributions to unions by individuals or corporations.

DUALISM IN THE AMERICAN LABOUR MOVEMENT

34. Examples of dualism. Dualism had reared its head several times in the history of American labour movements:

(*a*) *The Knights of Labour*. This organisation had a greater interest in political and social action than had the A.F.L. leadership in the 1880s.

(*b*) *Division of the A.F.L.* In the early period of its development the A.F.L. contained within itself a sharp division of opinion. On the one hand the socialists favoured more political action and an effort was made to move the A.F.L. on a more militant path. The Gompers school of thought favoured a line of approach of a more moderate nature, working with the capitalist system rather than against it.

(*c*) *Rise of the I.W.W.* This development threatened to divide the American labour movement along two diverging paths.

35. Formation of the C.I.O. In 1934 unions from within the A.F.L., dissatisfied with the executive's attempts to cater for workers in the most productive industries, formed the Committee for Industrial Organisation to expand union membership on an industrial rather than a craft basis in the automobile, rubber, steel, aluminium and radio industries. The aim was to unite skilled, unskilled and white-collar workers into industrial unions and to use such forces to compel employers in hitherto anti-union areas of industry to recognise labour's right to bargain collectively. Militant action was deemed essential for this to be done.

The executive of the A.F.L. promptly suspended the unions concerned, which coalesced under the leadership of John L. Lewis of the United Mineworkers.

36. Success of the C.I.O.

(a) The C.I.O. *had rapid success*. By 1937 it claimed a membership greater than that of the A.F.L. and by 1942 it had a membership of 5 million. In 1946 both the A.F.L. and C.I.O. each had a total membership of 6 million – about 10 per cent of the total labour force in each, the total labour force being approximately 61 million in that year.

(b) In 1938 the C.I.O. *adopted its own constitution* and changed its name to Congress of Industrial Organisation.

(c) A *militant policy* was followed:

(i) In 1937 the General Motors and Chrysler Corporation recognised the C.I.O. for collective bargaining purposes. Unions were recognised elsewhere in the automobile industry and in 1941 the last bastion, the Ford Motor Company, succumbed.

(ii) The United States Steel Corporation changed its long-held beliefs and recognised the C.I.O. and the rights of workers to belong to unions in 1937.

(iii) In 1946 coal, steel and vehicle production were badly affected by strikes. The union alienated public opinion and the way was paved for the passing of the *Taft–Hartley Act* to give the state more control on the action of unions. In the short space of fifteen years, unions had grown from a position of relative weakness into one of such strength that need was felt for a curbing of their power.

PROGRESS TEST 14

1. Discuss the growth of American trade unions before 1914. (**1–3, 5–17**)

2. Outline the origins and the subsequent growth of the A.F.L. (**15, 17–22**)

3. To what extent did the law frustrate trade-union growth in the U.S.A.? (**25–29**)

4. How important was the *National Labour Relations Act*, 1935, in trade-union affairs? (**31**)

5. Why was the *Taft–Hartley Act*, 1947, passed? (**33**)

6. Why was the C.I.O. successful? (**34–36**)

PART SEVEN

FOREIGN TRADE

CHAPTER XV

THE INTERNATIONAL ECONOMY

THE INTERNATIONAL ECONOMY BEFORE 1914

1. International trade, 1850–1914. During the nineteenth century a complex structure of international trading relationships evolved as the number of commodities and goods in international trade increased and nations became more interdependent.

Industrialisation induced a growth of international trade. Producers sought further afield for raw materials when indigenous supplies were inadequate, and sought markets for manufactured goods. A complicated pattern of multilateral trade was built up.

2. Factors in the growth of international trade.

(*a*) *Transport innovations* reduced overland and intercontinental freight and passenger costs and facilitated the opening up of new low-cost primary producing areas. Bulk freight transport became an economic proposition.

(*i*) After 1870 technical improvements in steamships reduced water-borne transport costs to the extent that some domestic producers found themselves unable to compete in their own markets against foreign producers; *e.g.* Britain's grain producers could not compete successfully in some sections of the British grain market against American producers despite the costs of transporting grain from the Prairies to Britain.

(*ii*) The extension of railway facilities and cost-reducing improvements facilitated the opening up of hitherto remote areas to trade in the large land masses. By 1900 more than 400,000 miles (64,000 km) of railway track had been constructed – nearly 50 per cent of this in the U.S.A., where rail charges fell by 60 per cent between 1864 and 1900.

(*b*) *Commercial policies* in Europe to the late 1870s and in the U.S.A. to 1864 became more liberal. (After 1879 protectionism revived in Germany and France and the American tariff became more protectionist.) Britain became a free-trade nation and did not restore protective tariffs before 1915.

(c) *Population* increased rapidly in Europe and the U.S.A., creating larger markets and providing the necessary labour supplies for further industrialisation. Real incomes tended to rise: markets increased in depth as well as width, particularly in the U.S.A.

(d) *Communications* technology advanced so that producers and buyers in different parts of the world could contact each other with relative ease.

(e) *Refrigeration* made possible the long-distance transportation of perishable goods.

(f) *Migration* of population occurred on a vast scale. Between 1840 and 1920 more than 25 million people emigrated to the U.S.A.

(g) Massive flows of *capital* – from Britain, the U.S.A., Germany, France, etc. – assisted development in various parts of the world. British capital was of importance in the early stages of economic development in the U.S.A., but by 1914 the U.S.A. was herself an exporter of capital. By 1914 Britain, the U.S.A., Belgium, France, Germany and Holland had investments abroad totalling nearly $10 billion – about 40 per cent British capital.

3. The "Atlantic economy." Significant trading links between the U.S.A. and Britain were forged before 1860. The U.S.A. became a supplier of primary products to Britain and absorbed significant quantities of British manufactures. Between 1840 and 1860 the U.S.A. supplied nearly 25 per cent of Britain's total imports. The U.S.A. also became an important market for British manufactures – especially woollens, linens, and metal manufactures – and became a good market for cotton goods. Britain's imports from the U.S.A. consisted of raw cotton, grain, meat and tobacco.

THE INTERNATIONAL ECONOMY, 1914–50

4. Upheaval. Two world wars, a severe economic depression in the years 1929–32 and economic nationalism in the inter-war period changed the international economy drastically.

(a) *The First World War.* This war affected Britain's trading position. She lost markets in textiles to overseas producers and she had to sacrifice some of her overseas investments to help pay for the war. She remained a creditor on capital account, but the U.S.A. became the largest creditor nation. In the inter-war period Britain's main export industries – steel, coal, cotton and shipbuilding – were in difficulties: her exports failed to reach the 1913 level, even in the 1920s.

(b) *Trade in the 1930s.* Economic nationalism, international monetary difficulties and the 1929–32 depression killed off the growth in international trade. World trade in the 1930s did not reach the level of 1929.

(c) *Control of immigration.* International migrations of population came under increasing control. The free flow of labour into the U.S.A. was severely cut back by the imposition of the literacy tests and immigration quotas in 1917 and 1921.

(d) *The gold standard.* The gold standard had worked with reasonable efficiency before 1914 in enabling nations to effect multilateral settlements, but it failed to work satisfactorily in the inter-war period. After 1930 nations resorted to managed currencies in efforts to clear up unemployment difficulties – the gold standard was abandoned.

(e) *Protectionism.* After the 1929 collapse of the American economy protectionism increased. Britain substantially abandoned free trade in 1932 to protect industries which had failed to adapt to new market conditions. In 1930 the U.S.A. increased her wall of protective tariffs and resorted in the 1930s to reciprocal trade agreements. Trade tended to become bilateral instead of multilateral. The state became more and more influential in economic affairs.

(f) *The Second World War.* This war caused further changes in international economic relationships. The U.S.A. became easily the most powerful economic force in the international economy: she was the largest creditor nation. In 1919 Britain experienced balance of payments difficulties for the first time since 1847 and had to face more acute difficulties between 1931 and 1947. Two world wars had robbed her of much economic vitality and she had to lean heavily on the U.S.A. for assistance, as did other western European countries. The wheel had turned full circle: in 1850 capital flowed into the U.S.A. from Europe to help an emergent nation to develop; in 1950 American capital helped restore a shattered Europe under the Marshall Aid plan.

PROGRESS TEST 15

1. What were the factors involved in the growth of international trade before 1914? (2)

2. How significant was the "Atlantic economy"? (3)

3. How did the First World War upset the international economy? (4)

4. Discuss the effects of the Second World War on international economic relations. (4)

CHAPTER XVI

BRITISH FOREIGN TRADE

INTRODUCTION

1. The eighteenth-century pattern.

(a) *1700–1750*. Britain was largely self-sufficient in raw materials and foodstuffs in the first half of the century. In fact exports consisted mainly of surplus indigenous raw materials and products manufactured from them – coal, lead, tin, woollen goods, woollen yarn, etc.

(b) *1750–1800*. In the second half of the century the pattern began to change. Imports of raw materials became more significant; exports of goods manufactured from imported raw materials (*e.g.* cotton goods) increased in importance. In years of dearth there was a marginal dependence on imports of grain.

2. Changes, 1800–50.

(a) *The growth of industry*. The economy became more industrialised and more orientated towards overseas trade. Reliance on imports of raw materials increased and exports were dominated by cotton textile manufactures made from imported raw materials: more than 60 per cent of exports were textiles by 1850. Indigenous products became less important in the export picture.

(b) *Agriculture*. Agriculture was still the major single industry. but imports of foodstuffs were of increasing significance. The industry did not increase its output at the rate that the population grew.

STRUCTURE OF OVERSEAS TRADE, 1850–1914

3. Importance of indigenous resources.

(a) *Raw materials*. In 1850 home supplies of wool, lead, tin, copper, iron, flax and coal provided the bulk of the raw materials used in the respective industries: the major imported raw material was cotton. By 1913 indigenous supplies of raw materials for industry were of minor significance except for coal: 80 per cent of raw wool used was imported; 90 per cent of lead, 80 per cent of tin, 80 per cent of flax and almost all the copper came from abroad.

(b) *Coal.* Coal grew in significance as a raw material for consumption by the nation and for exporting: in 1850 3 million tons were exported, valued at £1·3 million and less than 2 per cent of exports in value; in 1913 73 million tons were exported (plus 25 million tons of bunker coal in foreign ships), valued at £54 million – about 10 per cent of total exports in value.

(c) *Foodstuffs.* The main source of foodstuffs in 1850 was British agriculture. Caird, in 1867, estimated that Britain's farmers supplied 75 per cent of all foodstuffs consumed in the country. By 1900 the proportion was down to 50 per cent and by 1913 more than 55 per cent of foodstuffs were imported. Of the total imports of all kinds more than 40 per cent consisted of foodstuffs in 1900.

4. Character of imports. The composition of the total import picture changed greatly between 1850 and 1914:

(a) *1850.* Raw cotton was of greatest importance at the beginning of the period, the official value being £21 million. Grain imports totalled £12 million and sugar £10 million. Raw materials made up the bulk of the value of imports.

(b) *1890.* By this year grain and flour had surged to the top of the import list: raw cotton was second in importance, followed by raw wool. Meat, animals, sugar, butter and margarine were high in the list in terms of value.

(c) *1913.* Grain and flour imports were valued at more than £80 million in 1913, raw cotton at £70 million, meat and animals at £57 million, raw wool at £35·6 million and timber at £34 million. Other imports valued at more than £20 million were butter and margarine, sugar, non-ferrous metals and manufactures, rubber and oils, oil seed, etc. Valued at more than £10 million were tea, silk yarn and goods, iron and steel, hides, skins and furs, and petroleum products.

5. Sources of imports.

(a) The U.S.A. in the period as a whole was the most important supply region, particularly for raw cotton, wheat, meat and meat products. Significant quantities of wheat came from Prussia/Germany until the 1880s, from Russia, Canada (after 1860), the Argentine (after 1886), India (from 1874) and Australia (from

1873). After 1903 in various years supplies of wheat from Russia, Canada, Argentina, Australia and India each exceeded supplies from the U.S.A., which was the main supplier from 1873 to 1903.

(b) Germany successfully invaded Britain's domestic market as a supplier of iron and steel and dyestuffs from the late nineteenth century onwards, and the U.S.A. was exporting iron and steel manufactures to Britain by 1899.

(c) At the turn of the century, the U.S.A. became a less attractive bargain counter for foodstuffs, and Canada began to take over some of the British market. Germany supplied meat and dairy produce in greater quantities after 1880.

(d) Supplies of wool came largely from Australia, New Zealand and South Africa.

6. Exports.

(a) *The character of exports*. This underwent a radical change:

(i) Textiles lost much of their proportionate importance, making up only 25 per cent in value of all exports in 1913 compared with 60 per cent in 1850. As a proportion of exports of manufactures the decline was from 70 per cent to 50 per cent.

(ii) Cotton goods were the predominant textile exports and were more than double all other textile exports in value in 1913.

(iii) Coal increased in significance: by 1913 it made up about 9 per cent of total exports.

(iv) Iron and steel exports were increasing up to 1913, but imports of iron and steel (£15·2 million) must be offset against the exports (£55·4 million).

(v) Chemicals and machinery were of increasing importance, and from 1899 new ships and boats were exports of significance.

(b) *Markets for exports*. Here also there was change:

(i) Main textile markets in the years 1840–50 were Europe, the U.S.A., Argentina and Brazil, but in 1913 they were India, China and Japan. There had been a shift from areas of increasing competition for British exporters to less developed areas.

(ii) Europe, however, remained an important market because it absorbed considerable quantities of coal, machinery and re-exports such as raw wool and raw cotton.

(iii) The American market continued to be important, but colonial areas absorbed more significant quantities of exports and British business activity tended to be influenced more by international business trends than solely by American business trends.

(c) *Re-exports.* In 1913 the value of re-exports was in excess of 10 per cent of the total value of exports. Commodities of particular importance were raw wool (which went to Europe and the U.S.A.), rubber (to Europe and the U.S.A.), raw cotton and hides, skins and furs. However, as a proportion of total exports, re-exports declined in importance from 1880, when they reached 16 per cent of the total.

(d) *Fluctuations.* Increases in exports were not at a steady rate: there were considerable fluctuations. In short periods the *volume* declined from time to time, *e.g.* 1860–2, 1875–6, 1884–5, 1890–3, 1899–1901 and 1907–8. Between 1850 and 1913 the volume of exports and re-exports increased by nearly 700 per cent.

(e) *Terms of trade.* The amount of imports that one unit of exports would purchase varied. Long-term trends were; (i) 1819–57 unfavourable; (ii) 1860–73 favourable; (iii) 1874–81/4 unfavourable; (iv) 1885–1913 favourable. There were fluctuations within these periods.

OVERSEAS TRADE AFTER 1914

7. The inter-war period. Exports never attained the 1913 volume in the 1920s and were reduced to 55 per cent of the 1913 total in 1931: in 1938 the proportion was about 70 per cent. Britain's share of world exports diminished, too.

(a) *The reasons for the changes* were as follows:

(i) The loss of textile markets to Japan and India.

(ii) The loss of coal markets to lower-cost producers in Europe.

(iii) The loss of the iron and steel export market because of high costs in the British industry.

(iv) A fall in total world trade during the depression 1929–32.

(v) Increasing economic nationalism and a raising of tariff barriers.

(vi) The tying of the pound to gold at too high a parity in the 1920s and inept government monetary policies.

(vii) The loss of markets for shipbuilders because of the recession in trade.

(viii) A failure to change over rapidly enough to "new" industries to compensate for difficulties in the staple industries.

(b) *Imports increased in volume*, but this was compensated for, to some extent, by a favourable shift in the terms of trade so that for some years in the 1930s the value of imports was less than the 1913 total.

(c) *The character of exports and imports changed.* Exports of electrical goods, machinery, vehicles and aircraft increased in value: coal, textiles, and iron and steel decreased in value but remained important categories. Imports of manufactures, fuels and foodstuffs increased.

(d) *Export markets changed in relative importance.* By 1938 India and the rest of Asia, the U.S.A., South America and western Europe had reduced by significant amounts the value of British exports they absorbed. These were areas where competition proved fierce in the 1920s. Africa and Empire countries became relatively more important market areas.

8. The years 1945-50. The Second World War exacerbated Britain's position in the international economy and left her with acute balance of payments problems, necessitating a huge expansion of exports after the war, helped by Marshall Aid. There was a concentration on building up exports in expanding fields of trade, and by 1950 the trade gap was reduced to a mere £50 million.

FREE TRADE AND FAIR TRADE

9. Free-trade policy in the nineteenth century.

(a) *Industrial interests in Lancashire* formed an effective free-trade pressure group in the 1830s. Cotton manufacturers were interested in removing tariffs on imports and exports where possible because:

(i) all raw cotton was imported;

(ii) duty on grain maintained grain and bread prices at higher levels and indirectly raised the lower limit on wages in industry, thereby keeping costs higher;

(iii) it was argued that tariff reductions on imports would lead to reciprocity abroad and enable exporters to increase their sales.

(b) *Sir Henry Parnell* in 1830 recommended:

(i) simplification of the tariff system;

(ii) abolition of import prohibitions and duties on exports;

(iii) removal of duties on imports of raw materials, foodstuffs and manufactures;

(iv) retention of a few duties on articles in common consumption for revenue purposes only.

(c) *Free trade emerged slowly,* largely as a result of:

(i) the relaxation of duties and prohibitions by Canning and Huskisson in the years 1822-6;

(*ii*) the customs reforms of Peel in 1842 and 1845–6, and the repeal of the Navigation Laws in 1849;

(*iii*) the tariff reforms of Gladstone in 1853 and 1860 (including the Anglo-French treaty: *see* **11** below).

10. The Corn Laws. Controversy raged about the retention of tariffs to protect landowners and farmers. It became intense in the 1830s and 1940s:

(*a*) In 1838 the *Anti-Corn Law League* was inaugurated to stimulate discussion about the repeal of the Corn Laws and to pressurise Parliament into taking action. Notable among participants in the debates were Cobden, Bright, Bowring, Hume and McGregor.

(*b*) In 1846 *Peel* persuaded Parliament to repeal the Corn Laws.

11. The Cobden–Chevalier Treaty. This treaty between France and Britain was signed in 1860 to liberalise trade between the two countries and contained a "most favoured nation" clause.

Between 1862 and 1867 France signed treaties with all major European countries except Russia, and each of these countries negotiated treaties with each other and with Britain, use being made of the "most favoured nation" clause in each case. By 1867 France had enabled a network of treaties to emerge which reduced tariff barriers in Europe.

12. Agitation for fair trade. Tariff barriers began to rise again in Europe after 1878, partly in response to the need to protect agriculture against cheap American grain and partly to protect industry. Germany, France, Italy and Austria raised their tariffs. America in 1890 and 1897 increased her tariff wall.

A revision of British tariffs was demanded between 1881 and 1906, when the issue was temporarily decided in favour of free trade. Tariff changes were requested on the following grounds:

(*a*) Some industrialists desired the raising of tariff barriers against foreign manufacturers to minimise competition in the British domestic market.

(*b*) It was considered unrealistic for Britain to remain a free-trade nation when the U.S.A., Germany and France were increasing tariff barriers.

(*c*) The raising of tariffs would enable Britain to apply preferential duties towards goods from the colonies (Joseph

Chamberlain and the Tariff Reform League were in favour of this), and provide a lever for forcing down overseas tariff barriers and increase British penetration of European markets.

THE CHANGE FROM FREE TRADE TO PROTECTIVE TARIFFS

13. The years 1915–31. Before 1932 minor amendments to commercial policy resulted in some degree of protection for British industry:

(a) In 1915 the *McKenna Duties* were imposed:

(i) Duties were imposed at $33\frac{1}{3}$ per cent *ad valorem* on motor cars, cycles, watches, etc., and the importing of dyestuffs, chemicals etc., was subject to obtaining a licence.

(ii) The duties were imposed to save shipping space and foreign exchange and were retained after the war.

(b) In 1920 the *Dyestuffs (Import Regulations) Act* and *Safeguarding of Industries Act* were passed:

(i) Imports of dyestuff could be made only under licence.

(ii) The $33\frac{1}{3}$ per cent duty on imports was extended to glassware, pottery, cutlery, etc., and retained on goods in the McKenna list.

(c) In 1925 duties on *silk* and *hops* were imposed, and a *sugar-beet* subsidy granted.

(d) In 1927 the *film industry* was protected.

14. Protective tariffs, 1931–9. The collapse of the American economy after 1929 and the raising of tariff barriers in 1930 (*Hawley Smoot Act*) by the U.S.A. threw the international economy out of gear.

(a) The *Abnormal Importations (Customs Duties) Act*, 1931, permitted the imposing of tariffs whilst the country resolved its future policies.

(b) The *Horticultural Products Act*, 1931, extended duties to barley, oats and horticultural produce.

(c) The *Import Duties Act*, 1932, enabled a duty of 10 per cent to be applied until the Import Duties Advisory Committee reported. The Committee's report was followed by the imposing of higher duties at rates that varied according to the assumed need for protection, *e.g.* manufactured goods, 20 per cent duty; bicycles and chemicals, $33\frac{1}{3}$ per cent duty.

(*d*) The *Wheat Act*, 1932, allowed wheat producers to be subsidised.

15. Results of protective tariff policies.

(*a*) A policy of *imperial preference* was evolved.

(*b*) Britain had *bargaining weapons* whilst negotiating trade agreements. Twenty bilateral treaties were signed.

(*c*) *A sharp reduction in imports* occurred which eased balance of payments difficulties.

(*d*) *The difficulties of the staple industries were eased somewhat*.

(*e*) There was a *net increase in trade* with some transfer of trade from foreign to Empire markets.

16. Other policy aspects of the 1930s.

(*a*) *The abandonment of the gold standard in 1931*, and the depreciation of the pound sterling, was an effective protective measure until other countries devalued.

(*b*) *Agricultural marketing boards* were set up to organise the marketing of produce.

17. Difficulties and policies after the Second World War.
Europe's industry was badly mauled in the Second World War and required liberal trading conditions if recovery was to be reasonable in magnitude and speed.

The United Nations failed to erect an acceptable international trade organisation, and so the U.S.A. used its *Reciprocal Trade Agreements Act* as a base for negotiating the *General Agreement on Tariffs and Trade* with Britain and other European countries.

THE BALANCE OF TRADE AND THE BALANCE OF PAYMENTS

18. Adverse balance of trade.
Throughout the hundred years 1850–1950 the balance of trade was adverse. The income from exports was less than the expenditure on imports.

19. Balance of payments.

(*a*) A *favourable balance* was recorded every year from 1848 to 1919. The deficit in the balance of trade was more than compensated for by:

(*i*) the income from investments abroad;

(*ii*) the earnings from shipping, insurance, banking, etc. (these are referred to as *invisible exports*).

(*b*) *Balance of payments difficulties* increased after 1925. A negative balance was recorded in 1926, 1931–4, 1936–47 and 1949 because of a fall-off in world trade in the 1930s and in the Second World War, which reduced earnings from shipping, insurance, etc., and because British assets abroad had to be realised to pay for the two world wars. The situation was desperate by the end of the Second World War and a massive expansion of exports was necessary. The revival of exports was so successful, aided by a favourable shift in the terms of trade, that comfortable surpluses were achieved in 1950, 1952–4 and 1956–9.

PROGRESS TEST 16

1. Analyse the pattern of British foreign trade before 1850. (**1, 2**)

2. What were the main imports and from which sources did they come in the period 1850 to 1914? (**4, 5**)

3. Examine changes in the character of exports before 1914. (**6**)

4. Which were the chief markets for exports and re-exports before the First World War? (**6**)

5. Discuss the changes in the structure of British overseas trade after 1918. (**7, 8**)

6. Account for the balance of payments problems that grew after 1918. (**7, 8, 19**)

7. Outline the development of free trade in the nineteenth century. (**9–12**)

8. Why did Britain's commercial policy change after 1914? (**13, 17**)

CHAPTER XVII

AMERICAN FOREIGN TRADE

THE GROWTH OF TRADE BEFORE 1860

1. Rates of growth. American foreign trade fluctuated in its rates of growth from 1815 to 1860. Periods of rapid growth were 1815–18, 1831–7 and 1846–57.

(a) 1815–18. Trade grew rapidly because of the expansion of the British cotton industry and the demand for raw cotton, and because of British-financed trade in cotton manufactures. Concern was expressed about the growth of imports of manufactures from Britain.

(b) 1831–7. Tariffs were reduced in 1832 and stimulated trade. British capital helped finance transport improvements, and the cotton producers were subject to increasing demand for raw cotton to feed the Lancashire mills. As in 1818 the boom was followed by a severe recession and a repudiation of debts.

(c) 1846–57. Imports and exports more than trebled in value. European demand for American goods increased. Raw cotton dominated exports, but other agricultural products were of increasing significance and by 1860 manufactured goods were gaining in importance – especially cotton manufactures.

The rate of growth of foreign trade was slightly more than the British rate of growth of such trade from 1816 to 1860 but its relative importance in the economy was not so great as in Britain.

2. Distribution of trade.

(a) Links with *Britain* were strong:

(i) Between 1821 and 1860 Britain absorbed more than 45 per cent of American exports and supplied nearly 40 per cent of imports.

(ii) Britain was the chief buyer of raw cotton The combined French and German trade was much less than the trade with Britain.

(iii) Imports from Britain were cotton manufactures (many for re-export), iron rails, woollens, linens, silk, etc. American textiles began to replace British cotton manufactures by 1840.

(b) *France* and the *West Indies* were next in importance, although the West Indies declined as a market for exports in the 1850s. Sugar was imported from the West Indies, silk and woollens from France.

(c) *Europe* absorbed large quantities of re-exports – sugar, coffee, cocoa, dyewoods, etc. In 1860 Europe took about 75 per cent of American exports.

(d) *New York, Boston, New Orleans, Philadelphia* and *Baltimore* were the main ports associated with overseas trade. New York was of predominant importance.

OVERSEAS TRADE, 1860–1914

3. Fundamental changes.

(a) The *industrial and agrarian revolutions* gathered pace in this period so that in addition to being the premier producer of cotton the U.S.A. became the leading producer of the following commodities:

    (i) Grain and livestock.

    (ii) Iron ore, coal, copper and oil.

    (iii) Iron and steel.

    (iv) Motor vehicles.

(b) Changes in the *import–export pattern* occurred:

    (i) There was a proportionate increase in the export of finished manufactures and semi-manufactures. Exports of crude materials, crude foodstuffs, processed foodstuffs diminished relatively.

    (ii) Imports of crude materials, semi-manufactures and processed foodstuffs increased relatively and absolutely. Imports of finished manufactures declined from 50 per cent to about 20 per cent of total imports.

(c) Between 1860 and 1920 the *value of exports* at current prices increased more than twenty-five times and the value of imports about fifteen times. The balance of trade changed from adverse in 1860 to favourable by 1875. The U.S.A. became a creditor nation on current account.

4. The impact of the extension of the frontier.

The tremendous increase in the acreage under the plough after 1860, plus the cheapening of methods of production and transport, had tremendous international effects:

(a) The U.S.A. served as a cheap source of food for Britain

especially, and for other European countries, between 1875 and 1900 and, to a lesser extent, from 1901 to 1914.

(b) British grain farmers found that American grain producers could sell more cheaply in British markets than they could themselves.

(c) Germany and France raised tariff barriers to keep out cheap American agricultural produce.

(d) Outpouring of agricultural exports helped build up a favourable balance of trade for the U.S.A.

5. The impact of industrialisation.

(a) *Manufactures.* The rate of growth of output of manufactures was such that by 1914 exports of manufactures were of similar magnitude to exports of agricultural produce and by 1920 much greater in value. Exports consisted of iron and steel products such as iron and steel plates, office equipment, sewing machines, loco-motives, electrical equipment, textiles, copper products, etc.

(b) *Markets.* Markets changed somewhat. Europe declined relatively (but not absolutely) as a purchaser of American exports, and Asia, Africa, Canada, Central and South America were devel-oped as market areas by American firms which established their own selling agencies and catered for specific foreign requirements. (Europe still absorbed more than 60 per cent of American exports in 1913–14.)

(c) *Raw materials.* Increasing concentration on manufactures implied an increasing demand for raw materials, some of which (rubber, tin, wool, nickel, copper, jute, etc.) had to be imported. There was a substantial increase in imports from 1900 to 1914, particularly from Asia. The U.S.A. had ceased to be a primary source of raw materials for foreign producers.

OVERSEAS TRADE AFTER 1914

6. The impact of the First World War.

(a) *Exports* of foodstuffs declined both relatively and absolutely from 1900 to 1910, but the First World War altered this trend. Lower-cost areas of production such as Australia and South America temporarily lost European markets to the U.S.A. because of shipping difficulties. In the later part of the war foodstuffs exports increased both relatively and absolutely. Europe also absorbed larger quantities of American manufactures.

(b) The decline of Europe as *a supplier of American imports* was hastened by the war. Supplies of chemicals, textiles and other manufactures diminished and a healthy re-export trade with the U.S.A. was lost because of shipping scarcities. The American mercantile marine was enlarged and the country traded directly with original producers.

(c) The U.S.A.'s *debtor status* on capital account was rapidly extinguished as European capital invested in the U.S.A. was repatriated and loans were sought from the U.S.A.

7. Inter-war period, 1918–29.

(a) Despite restrictive tariff barriers and the imposition of quotas *world trade expanded* in the 1920s. American imports of raw materials increased: countries such as Brazil, Cuba, Malaya, Japan, the Dutch East Indies, etc., became somewhat dependent on the American market. Exports of agricultural products were high between 1918 and 1920 but declined as European farmers recovered from the war and as other low-cost areas re-entered into competition with American farmers in overseas markets. Manufactures continued to be the chief exports in value.

(b) *Capital* from the U.S.A. was needed by Europe to promote recovery after 1918. By 1929 direct investment by American firms abroad was necessary to act as a counter-balance to the U.S.A.'s position of creditor on both current and capital account.

(c) *Prospects of speculative profits* on the stock exchanges induced withdrawal of American capital from Europe in 1928 and 1929 and put the international economy under great stress. The collapse of the American economy towards the end of 1929 resulted in a collapse of the international economy.

8. Inter-war period, 1929–39.

(a) *The 1929–32 depression* cut back American foreign trade quite severely so that the volume of trade in 1932 was 70 per cent less than the volume in 1929.

(b) *Recovery* began in 1933, stimulated after 1934 by reciprocal trade agreements with twenty-six nations. Between 1932 and 1939 exports and imports roughly doubled, the greatest increase being in trade with agreement countries.

(c) *The Second World War* upset the international economy. American neutrality legislation had to be abandoned and a lend-lease agreement with Britain signed under which the Allies benefited by some $70 billion. In 1941 the U.S.A. was pulled into the conflict by the Japanese attack on Pearl Harbour.

(d) *After 1945* aid had to be poured into Europe to facilitate recovery. In 1948 the Marshall Plan was put into operation and by New Year's Day 1950 $39 billion of American capital had been poured into other countries since 1945. This helped offset the favourable balance of trade that the U.S.A. enjoyed.

COMMERCIAL POLICIES

9. Tariffs before 1860. Between 1816 and 1828 the policy of the U.S.A. towards overseas trade grew increasingly protective, against the wishes of the South. In 1830, 1832 and 1833 reductions of tariffs were effected, but in 1842 the trend was reversed, to be followed in 1846 by further reductions and in 1857 by significant reductions.

10. The growth of protection.

(a) The *liberal movement* in foreign trade was *checked* with the onset of the Civil War. Higher tariffs were imposed in 1861, 1862 and 1864.

(b) Some *10 per cent reduction* of duties was achieved in 1872 but the 1864 rates were restored in 1875 because of panic following the recession of 1873.

(c) In 1882 a large *Treasury surplus* existed and tariffs were reduced by 5 per cent (although an investigating commission had recommended a reduction of 25 per cent).

(d) The *McKinley Tariff* (1890) pushed duties to a new high level of 49·5 per cent. This tariff was much more protective than the German and French tariffs, although Congress did allow in free of duty sugar, coffee, tea and some raw materials if they were conveyed from the country of origin and if American goods received reciprocity.

11. Reasons for high rates of duty.

(a) *Pressure groups from industry* were able to persuade Congress from time to time to protect and succour industries by the imposition of import duties; *e.g.* the 125 per cent increase in duty on tinplate in 1890 fostered the rise of the tinplate industry in the U.S.A. and reduced imports from Wales dramatically.

(b) The *agricultural interest* was concerned about the growth of imports of produce and secured the extension of protection to agricultural products in 1890.

(c) High tariffs were regarded as protecting the high wages

paid to the *labour force*. One of the aspects of so-called "scientific tariffs" was that they reputedly equalised wage differences between the U.S.A. and foreign countries.

(*d*) *Federal income* was reaped mainly from tariffs in the nineteenth century. Proposals to replace tariffs by direct taxation would have been unpopular.

12. Pressure for lowering of tariffs. Before 1890 agrarian interests in the West and South agitated for a reduction of duties on imports of manufactures to keep down the cost of living in rural areas.

As manufacturing expanded and relied on imported raw materials to a greater extent so pressure was exerted for the reduction or abolition of duties on raw materials, and in 1894 the Wilson Tariff reduced duties on imports such as iron ore, coal, raw wool, etc.

13. The Dingley Tariff Act, 1897.

(*a*) The average level of duty was raised to 57 per cent.

(*b*) Provision was made for the negotiation of *reciprocal trade agreements* (as in 1890) and treaties with France, Italy and Portugal were ratified. The President was empowered to reduce duties by 20 per cent in negotiating such treaties, but the treaties were subject to ratification by the Senate.

14. Tariffs in the twentieth century.

(*a*) *Revision of the tariff* was sought in the early twentieth century and the first substantial change came in 1913 when some raw materials and agricultural products were freed of duty and cottons and woollens were subject to reduced duties. Other rates, however, were raised.

(*b*) A *Tariff Commission* was appointed in 1916 to examine tariffs and report to Congress – part of the search for a scientific tariff.

(*c*) After 1918 a liberal tariff policy might have been expected in view of the U.S.A.'s balance of payments position, but *protective* Acts were passed in 1921 (the *Emergency Tariff Act*) and 1922 (the *Fordney–McCumber* Act). These two Acts were passed largely to protect agriculture but distress in the industry was not prevented or alleviated. In 1930 the *Hawley–Smoot Act* raised the tariff to new heights and helped plunge the international economy into further distress.

15. The Reciprocal Trade Agreements Act and GATT.

(a) *Revival of foreign trade* was one of the planks of Roosevelt's administration and in 1934 the *Reciprocal Trade Agreements Act* empowered the President to negotiate trade agreements, authorising him to manipulate the Hawley–Smoot tariff rates by 50 per cent either up or down. These powers remained in existence by renewal until 1948 and 29 agreements were negotiated with a favourable impact on trade in the 1930s, reversing trends towards higher tariffs and greater restrictions on trade.

(b) *GATT* was evolved at a conference in Havana in 1948 and served as a basis for the liberalising of trade in the post-war world.

PROGRESS TEST 17

1. Discuss changes in American foreign trade between 1860 and 1914. (3–5)

2. How did the extension of the frontier affect American and European trade? (4)

3. How did the First World War affect American overseas trade? (6–8)

4. Explain how some recovery in American foreign trade was brought about after 1918. (7, 8, 15)

5. Examine the American tariff system. (9–15)

6. What was the significance of the *Reciprocal Trade Agreements Act*? (15)

PART EIGHT

BANKING

CHAPTER XVIII

BANKING IN BRITAIN

1. Introduction: economic growth and banking.

(a) As conditions for investment became more favourable in the eighteenth and nineteenth centuries so the *demand* for banking facilities expanded. A flow of capital from capital surplus areas to capital deficit areas was required: would-be borrowers required access to the capital of those able and willing to lend it. Banks provided a vital link between borrowers and lenders.

(b) During the seventeenth century in Britain the depositing of valuables with *goldsmiths* increased and provided a base from which banking could expand, firstly on a purely private basis, and then on a basis linked with the Crown's needs for finance. Private banks grew mainly to serve particular areas, but the *Bank of England* was formed in 1694 to supply funds for the Crown, and enjoyed a monopoly of joint-stock banking in England and Wales until 1826. The needs of industry and commerce were served by a rapidly expanding private banking system – by 1821 about 800 private banks existed. In Scotland banking developed more soundly on joint-stock lines.

(c) Whilst early British banks arose through *private initiative* out of normal business activities, early American banks were formed under *state charters* to facilitate borrowing by the state governments and were modelled on the Bank of England.

(d) During the nineteenth century banking in Britain gradually adopted the *joint-stock* form, and with the Bank of England assuming the real role of a central bank, considerable safety and stability were achieved. In the U.S.A., however, between 1836 and 1864 no federal system existed within which stability and safety could be achieved. Whilst the power of note issue in Britain was concentrated in fewer and fewer hands, in the U.S.A., to 1863, a multiplicity of note issues arose to give a confused state of affairs in banking. From 1864, however, the *National Bank Act* was a corrective influence, but, even under the Federal Reserve System, banking stability in the U.S.A. left much to be desired. National appetite for bank credit in the U.S.A. appears to have given rise to an expansion of bank enterprises on unsound bases that were liable to topple over in times of stress. Hence, in 1929–33 bank failures were widespread. However, the banking system in the U.S.A., because of the number of states involved and the size of the nation, is more complex than in Britain. Additionally, the system in Britain evolved gradually from about the seventeenth century onwards; it had more than a hundred years in which to evolve before the American nation was formed.

EARLY PRIVATE BANKS AND JOINT-STOCK BANKS

2. Origins of private banks. The early private banks developed from business activities:

(a) Goldsmiths and silversmiths, operating mainly in London, began to accept deposits of money and valuables. From this banking was a natural development.

(b) Money scriveners acted as financial intermediaries and soon launched out as bankers.

(c) Industrialists, or manufacturers, developed banks:

(i) to ease currency shortages in particular areas; and
(ii) to make profitable use of funds that they had available.

The early private banks provided a necessary easing of pressure on credit channels and fostered a speeding up of industrialisation by transferring funds from savers to investors. Larger units tended to be in London and ran accounts for the country banks that evolved outside the metropolis.

3. Private banks, 1760–1826. Economic growth increased and was sustained after 1760. Industrialisation was pursued, stimulated by a rising demand for manufactures. There was a rapid growth of towns in northern and Midland mining and manufacturing areas. Urgently needed currency and banking facilities were provided by the growth of country banks when drapers, merchants, brewers, ironmasters, etc., developed banks as auxiliary economic activities to foster the growth of their primary business interests. By 1760 more than 700 private banks existed.

Although they filled an economic need, country banks had serious weaknesses:

(*a*) By law the maximum number of partners in a private bank was limited to six: therefore the capital employed was often too small in relation to the commitments undertaken.

(*b*) Until 1808 notes could be issued without a licence and, after the 1808 Act, anyone who could pay the £30 licence fee and the stamp duty on notes could issue as many notes as he thought fit, provided the denomination was at least £1. Many banks issued notes without ensuring cash reserves were adequate; consequently in times of stress many banks had to stop payment.

(*c*) Assets were not spread widely over a variety of industries. Usually they were tied up in one industry and in one or two large firms.

(*d*) Most country bankers maintained accounts with London banks, but communications were not speedy and very little systematic attempt was made to ensure that there was a proper relationship between cash reserves and reserves held with London bankers.

(*e*) The Bank of England was a competitor in many fields of business – a privileged competitor that had a monopoly of the joint-stock organisation in banking in England and Wales. It was not an institution that stood at the centre of the banking world, regulating activities. Banking, then, had no co-ordinated policy in the country as a whole. Units operated in relative isolation and it was difficult for individual banks to withstand pressure in times of panic. Failure rates were high in times of crisis. Financial panic resulted from a chain reaction: the failure of one bank caused suspicion about the soundness of other banks and often brought about the closure of banks previously considered eminently sound.

4. London private banks and the Bank of England. The Bank of England, established in 1694, was dominant in the London money market since it was the bank of the government and, therefore, closely concerned with public finance. However, its private business remained relatively small until after 1797, and the Bank's advent in 1694 did not bring to an end the life of London banks established by goldsmiths, etc.

(*a*) One early result of the coming of the Bank of England was the driving out of circulation of the *note issues* of London private banks. A more general use of *cheques* developed slowly and in 1770 the London private banks established their London Clearing House.

(b) London bankers began to deposit excess balances with the Bank of England, which thereby became a type of *reserve bank* and at times of stress would grant accommodation to city bankers to tide them over their difficulties.

(c) *Stability* of city banks tended to be greater than country banks because of the relatively greater resources they possessed. The failure of a London bank could have serious consequences for country banks whose London agent it happened to be. London agents provided various services for country banks – collection of bills and documents, making payments, providing foreign and domestic investment facilities, etc. Poles of London were agents for forty-three country banks and the failure of Poles in 1825 involved the subsequent failure of numbers of country banks. In that year seventy-three banks in England and Wales suspended payments to customers.

5. Joint-stock banks.

(a) *Pressure for reform.* The crisis of 1825, the failure of thirty-seven banks and the suspension of payments by double that number helped to intensify pressure for legislative reform of banking.

(b) *Bank of England's monopoly.* The Bank of England had a monopoly of joint-stock banking with note issue; no other banking firm with more than six partners could issue notes. Scottish banking showed a strong contrast with English banking: the Bank of Scotland's monopoly expired in 1716 and strong joint-stock banks developed branches, enabling the banks to diversify their assets and to withstand periods of crisis more easily than their English counterparts.

6. The Bank Act, 1826.

(a) *Banks outside London.* Joint-stock banks (any number of partners) were permitted outside a sixty-five mile (104-km) radius from the centre of London. The Act placed limitations on business by such banks within London and did not limit liability.

(b) *The Bank of England.* The Bank was permitted to open branches in the provinces and opened eleven branches by 1833.

(c) *Spread of joint-stock banks.* Banks were formed on a joint-stock basis in Lancaster, Norwich, Huddersfield, etc., and by 1833 there were about fifty such banks in operation. In 1836 there were a hundred – mainly banks of issue. Thomas Joplin was a primary figure in this development.

7. The Bank Charter Act, 1833. In 1833 the charter of the Bank of England was due for renewal. Joplin asserted that the existing charter did not preclude the opening of non-issue joint-stock banks in the London area, which suffered from a lack of banking facilities. In the renewal of the charter in 1833 a clause was inserted permitting the opening of joint-stock banks (without issue rights) in London.

8. The significance of the 1826 and 1833 Acts.

(*a*) They provided for the spread of joint-stock banks, with unlimited liability, in the provinces and London and permitted the opening of branches. They made for stronger banks with risks spread more widely over firms and areas.

(*b*) The Bank of England was allowed to retain monopoly of note issue in London, but the Acts did little to bring the issue of notes under central control. In 1841 287 private banks and ninety-one joint-stock banks issued notes.

(*c*) The Acts of 1826 and 1833 doomed small private banks to *gradual elimination*. Private banks diminished from a total of nearly 800 in 1821 to less than fifty by 1913. The number of joint-stock banks reached a total of 120 by 1870 and then diminished to forty-three by 1913 because of amalgamations.

THE BANK CHARTER ACT, 1844

9. Problems of the 1830s. Two problems became more acute in the 1830s:

(*a*) *The dichotomy facing the Bank of England* in its roles as a private profit-making institution with responsibility to shareholders and, at the same time, as a bank of the government entailing responsibilities of a public nature meant a clash of private and public interests could and did cause conflict within the Bank of England.

(*b*) *The proliferation of note issue* made control of credit and currency very difficult. Tooke observed that country banks did not necessarily contract issues of notes when the reserves of the Bank of England fell to dangerously low levels and could, therefore, increase internal drains of gold away from the reserve at a time when external drains were giving rise to acute anxiety.

10. Attempted solution. To attempt to solve these problems Select Committees on joint-stock banks met in the 1830s and in 1840 and 1841 Select Committees on banks of issue were appointed. Two schools of thought emerged:

(a) The *banking school*, which wanted note issue to fluctuate in accordance with business needs. Banking-school protagonists argued that the self-interest of holders of notes and issuers of notes would provide a system of checks and balances over the volume of notes in circulation and that there was no justification for interfering with existing rights of issues. Gilbert and Stuckey supported these views.

(b) The *currency school*, which favoured a linkage of note issues with gold reserves so that a contraction of the latter would be accompanied by a contraction of note issues. The currency school thought the approach advocated by the banking school was too unrealistic because:

(i) competition among banks of issue made for over-issue;
(ii) it would not remove the causes of financial panic.

To prevent conflict between private profit-making aims and public responsibility for note issue it was advocated that the Bank of England have two departments, *viz.*

(a) a banking department to follow normal banking practice;
(b) an issue department concerned solely with the supply of currency in accordance with the reserves of gold.

11. The Bank Charter Act, 1844. Under this Act the Bank of England remained a private company and two departments were created in the Bank:

(a) The *Issue Department* (to publish weekly summaries). The issue was fixed as follows:

(i) £14 million fiduciary issue – pegged at this level.

(ii) Notes issued in accordance with gold and silver bullion reserves.

(iii) The fiduciary issue could be increased by two-thirds of the issue of banks forfeiting issue rights. (New banks of issue were prohibited.)

(iv) Issues of other banks were not to exceed the average issue of the twelve weeks preceding 27th April 1844 and banks had to make a monthly statement of issues.

(v) Notes to be convertible on demand – gold at £3 17s. 9d. an ounce.

(*b*) The *Banking Department*. Its functions were to be as follows:

    (*i*) To publish weekly accounts.

    (*ii*) To operate as before except for concessions to London joint-stock banks which could draw, accept or endorse bills of exchange.

It was hoped that by providing some form of control over note issue cyclical fluctuations and financial crises would be minimised. The crisis of 1837–9 was thought to have been exacerbated by the policies of the banks of issue and of the Bank of England.

12. The effectiveness of the 1844 Act.

(*a*) *Fears as to rigidity*. The Earl of Radnor, John Masterman, M.P., Norman Bosanquet and others feared that the Act was too rigid in its regulation of currency issue: there was a plea to make the rules of issue more elastic. S. J. Lloyd (later Lord Overstone), however, thought that early action to safeguard reserves could be taken by the Bank of England and the Act should, therefore, minimise the oscillations of the economy between boom and depression.

(*b*) *Recurrent crises*. Over the next twenty-two years three crises occurred (in 1847, 1857 and 1866) for which the rigidity of the *Bank Act* of 1844 was held to be partially responsible by contemporaries (*see* **13–15** below). Permission to suspend the Act in relation to the fiduciary issue had to be sought in each of the three years.

13. The 1847 crisis.

(*a*) *The harvest of 1846 was deficient*. Imports of grain were sought far and wide, leading to a drain of gold to pay for the imports.

(*b*) *Capital was tied up in railway promotions*. In the years 1844–8 more than 9,000 miles (14,400 km) of railway track were authorised and 1,300 miles (2,080 km) were actually opened to traffic. In January 1847 capital sums totalling £6·15 million were called on by railway companies. Large sums, too, were invested in railways in Europe.

(*c*) *The 1847 harvest was good*. Wheat prices fell rapidly, inflicting losses on merchants who had large stocks of imported wheat, purchased at high prices, on their hands.

(*d*) *Banks with close connections with imports felt the strain*. Seven private and four joint-stock banks suspended payments.

The Bank of England refused to provide support: advances on stock or Exchequer Bills ceased. The Governor and three directors of the Bank of England were involved in firms which failed.

(e) *Reserves were falling.* By 23rd October 1847 reserves in the Banking Department of the Bank of England were less than £2 million. Application was made to the government for the fiduciary limit on note issue to be exceeded: this was granted. Panic gradually subsided and a quick recovery set in aided by railways, the stimulus of gold-mining and gold discoveries, growing demand for capital goods abroad and increasing industrialisation at home.

14. The 1857 crisis. The 1857 crisis was the first really worldwide economic crisis in history; it affected the U.S.A., Britain, Central Europe and South America.

(a) *American railway companies* had difficulty in paying interest on loans, which led to foreclosure by bond-holders.

(b) *American banks* closely associated with railway development were in difficulty: 1,517 banks closed their doors.

(c) *5,000 American business firms* collapsed.

(d) *Serious repercussions* were felt in Britain: banks and firms involved in the Atlantic economy were under strain. The Liverpool Borough Bank failed, the Clydeside Joint Stock Bank tottered on the brink of closure and Overend & Gurney had to ask the Bank of England for unlimited assistance should it be needed.

(e) There was an *external drain on bullion* in the years 1852–6 (partly due to the Crimean War) followed by an internal drain in 1857. By 9th November 1857 reserves of the Issue Department of the Bank of England were down to £581,000 and the fiduciary issue had to be exceeded by £2 million, of which £928,000 was put into circulation.

The financial crisis was soon over but business confidence was restored only very slowly: 1858 was one of the worst years of the nineteenth century for unemployment.

15. The 1866 crisis. This is largely the story of the collapse of Overend & Gurney, which had half London's discount business in its hands and unwisely lent too much on long term.

(a) The failure of Overend & Gurney, stated *The Times* of 11th May 1866, was a *national calamity* which destroyed business confidence. Liquidity preference rose. The Bank of

England had huge demands on its resources and reserves fell quickly.

(b) On 11th May the Bank of England received permission to *exceed the fiduciary limit*.

16. Failure of the Bank Charter Act. The Act failed to stop economic crises: it could not do that anyhow. One of the big failures of the 1844 legislators was that they did not recognise bank deposits and bills of exchange as money: there was confusion over what constituted money, so that in times of pressure the liquidity position of the Bank of England became somewhat parlous. However, in all three crises the difficulty over liquidity seems to have arisen because of a run on Bank of England notes and not on gold. The 1844 Act appears to have created greater confidence in banknotes than existed previously.

THE GROWTH OF JOINT-STOCK BANKS

17. Effects of legislation. The *Bank Charter Act*, 1844, and the *Registration Act*, 1844, had the twofold effect of slowing down amalgamation of banks at first because of unwillingness to forfeit issue rights and of enabling existing joint-stock banks to consolidate their positions because of a mild discouragement of new, large-scale ventures in banking.

18. Increased demand on banks. Economic and technological progress increased demands on individual banks and implied that the joint-stock form of organisation was likely to be that required in the future. In 1858 limited liability became available to banks merely by complying with the requirements of the Act. Growth in the size of bank units was thereby facilitated. In 1866 there were 246 private banks with 376 branches and 154 joint-stock banks with 850 branches. The number of banks declined as amalgamations took place but the number of bank offices increased, making banking facilities available to a greater number of people. Measured per head of population banking facilities more than doubled between 1870 and 1914. Total deposits with joint-stock banks doubled, too.

Amalgamation of banks was beneficial to the nation for the following reasons:

(a) Bank resources and risks were spread widely.

(b) Surplus funds could easily be transferred from one area to another.

(c) Larger banks could follow policies independently of a large customer's wishes.

(d) Substantial customers could be accommodated easily—a very necessary development as capital requirements of firms grew.

(e) Banking became safe because of amalgamations and because of co-operation among banks. Crises in 1878, 1890 and 1907 were weathered fairly easily because of the size and strength of the Bank of England and other bank units.

(f) The development of the "safe" bank system was of fundamental importance in the increasing stature of London as a leading centre of finance in the international economy.

19. The Big Five. Long years of amalgamation culminated in the emergence of the "Big Five" – joint-stock banks which, in 1920, jointly controlled 83 per cent of all bank deposits – and signalled the end of large-scale amalgamations. Growth after 1920 was largely one of filling in gaps in territorial coverage by the opening of more branch offices.

THE BANK OF ENGLAND AS CENTRAL BANK

20. Changing role of the Bank of England. The Bank remained a private company until 1946 but gradually donned the garb of a central bank in the nineteenth century. Bank rate began to be used as a weapon to act on the public demand for money rather than as a response to public demand. At first reserves were often inadequate to enable the Bank to fulfil its public role but reserves were accumulated after 1872 so that by 1896 they stood at £44 million.

(a) *Open market operations.* To make the Bank into an effective weapon the technique of open market operations was evolved to vary funds at the disposal of the money market. Three forms of this can be discerned in the history of the Bank – borrowing on consols, borrowing on the market, and borrowing directly from other banks.

(b) *Rising stature.* In 1890 the Bank came to the aid of Barings: Lidderdale enhanced the Bank's moral stature and convinced other banks that they should look to the Bank of

England for a lead. Despite relatively small reserves the Bank was able to withstand a drain of gold to the U.S.A. in 1907, further enhancing its reputation.

(c) *Forfeiture of issue rights.* Other banks surrendered rights to issue currency as consolidation of banking progressed. The last forfeiture of issue rights occurred in 1921 when Fox, Fowler & Company amalgamated with Lloyds Bank. During the 1914–18 War the Treasury issued currency notes exceeding in total value £300 million. In 1923 the fiduciary issue of the Bank stood at £19·75 million, the maximum permitted by the *Bank Charter Act* of 1844. The amount was increased to £260 million in 1928 when the Bank of England and Treasury notes issued in the First World War were amalgamated under the control of the Bank, leaving currency solely in the hands of the Bank of England. The amount of the fiduciary issue was subject to manipulation after 1928 according to the needs of trade and the state of reserves, policy being dictated by the Treasury. Currency in circulation was £324 million in 1929: by 1950 it was more than £1,240 million.

21. Convertibility of currency. Since 1798 the currency has alternated between convertibility and inconvertibility:

(a) *1797–1821.* Bank of England currency was inconvertible because of the wars with France and internal threats of gold reserves.

(b) *1821–1914.* The gold standard was in operation. The system appeared to work fairly smoothly.

(c) *1914–25.* Currency was theoretically convertible: in practice it was not. The pound was restored to gold in 1925, after five years of deflation, at the old parity. This caused a continuation of governmental deflationary policy in the 1920s to hold the value of the pound.

(d) *1925–31.* Convertibility at the rate which Keynes estimated was 10 per cent too high resulted in competitive disability in export markets and encouraged the growth of imports, with adverse effects on the balance of trade, the balance of payments and employment.

(e) *1931 onwards.* Currency again became inconvertible. Sterling became a managed currency with an Exchange Equalisation Account, established in 1932, to achieve stability in foreign exchange rates. The pound depreciated rapidly with favourable effects on exports and the balance of payments.

In 1949 sterling was devalued by 30 per cent, to $2·80, because of economic difficulties caused by the Second World War. A further devaluation to $2·40 followed in 1967.

22. Little change in fundamental policy. With regard to home investment, the fundamental policy of British banks underwent little change before 1939. Direct investment in the share capital of industry was not a general policy, and loans to agriculture and industry were intended to be short-term, self-liquidating ones.

The policy was in marked contrast to that of American banks. The Macmillan Committee of 1931 noted the advantages for industry in the U.S.A. and Germany of close support from banks and financial groups: it noted in particular the need for bank support for small and medium-sized firms for which the City made inadequate provision – the Macmillan Gap. Some efforts, however, were made to build closer relationships between banks and industry:

(a) In 1929 the Securities Management Trust was formed.

(b) In 1930 the Bankers' Industrial Development was formed for providing loans to industry.

(c) In 1930 the Bank of England secured holdings in the United Dominions Trust finance group.

(d) In 1934 the United Dominions Trust formed a subsidiary – Credit for Industry – to fill the Macmillan Gap.

(e) In 1945 the Industrial and Commercial Finance Corporation and the Finance Corporation for Industry were formed – with powers to lend to industry and to borrow for this purpose.

PROGRESS TEST 18

1. Examine the growth of British private banks in the nineteenth century. (**2, 3, 4**)

2. What were the provisions of the 1844 *Bank Charter Act*? (**11**)

3. Why did economic crises occur in 1847, 1857 and 1866? (**13, 14, 15**)

4. Discuss the growth of English joint-stock banks. (**17, 18, 19**)

5. Analyse the development of the Bank of England as a central bank. (**20, 21**)

CHAPTER XIX

BANKING IN THE U.S.A.

INTRODUCTION

1. Britain and the U.S.A.: differences. Constitutional differences make more complex the study of American banking as compared with British banking.

(a) Britain is a unitary state, having one legislature – Parliament.

(b) The U.S.A. has a federal constitution and has a federal legislature and state legislatures, and, therefore, Acts relating to banking are passed by both. Banks formed under state Acts are referred to as state banks.

2. Chronological divisions. American banking history can be divided chronologically as follows:

(a) *1783–1837*. State and federal banks were incorporated by special charter – either a state charter or, in the cases of the two Banks of the U.S.A., 1791–1811 and 1816–36, special federal charters.

(b) *1837–63*. This was the period of "free" banking, the era of general laws of incorporation at state level enabling bank associations to be formed by compliance with such laws. It was also a period when banks faced much hostility in states such as Oregon, Texas, California and Arkansas after the collapse of banks in the 1837–42 recession.

(c) *1863–1913*. The National Bank period marked an attempt to bring some semblance of order to the American banking world under the *National Bank Act* of 1863–4. In particular there was an attempt to bring the issue of banknotes under control.

(d) *1915 onwards*. This was the Federal Reserve area. Further endeavours were made to control American banking to inject greater stability into the system.

AMERICAN BANKS, 1783–1837

3. The nature of early banks in the U.S.A. By 1790 there were four banks in existence:

(a) The Bank of Philadelphia (formed in 1782 and continuing to 1929).

(b) The Bank of New York (formed in 1784 and still in existence).

(c) The Massachusetts Bank of Boston (formed in 1784, it became a National Bank in 1865 and was absorbed by the First National Bank of Boston in 1903).

(d) The Bank of Maryland, Baltimore (formed in 1790 and continuing until 1834).

NOTE: Attempts had been made to set up banks in Providence, Richmond and Charleston.

Each bank operated under a state government charter and could be compared with the Bank of England in that each was instituted to facilitate the loans of funds to the state government as well as to conduct normal business. Banking, therefore, began with the formation of *public banks* which were modelled on the Bank of England.

4. Origins of banks. Whereas banks in England originated partly in response to a pressure of supply of capital and partly in response to the demand for capital, the American banks arose largely as a response to demands for capital that was relatively very scarce. As business activity expanded, with a consequent increase in demand for capital, as the population increased and as the number of public improvements increased so did the pressure on the banks for credit. The number of banks totalled almost 800 by 1837.

5. The First Bank in the U.S.A.

(a) *Hamilton's Bank.* The constitution of the U.S.A. was silent about the setting up of a federal bank. During the years 1783–91 the monetary situation in the U.S.A. was very confused; a variety of types of coinage of uncertain value, which fluctuated with the distance from the place of origin of the coinage, and state laws of great variety made the smooth working of commercial transactions difficult. Economic growth was retarded by the lack of a widely acceptable currency of stable value. In 1791 the First

Bank of the U.S.A. was chartered by Congress despite the opposition of Jefferson and his followers. Hamilton's Bank was to be situated in Philadelphia and was to serve the federal government in the same way that the Bank of England served the British government, although, essentially, it was a private commercial bank with a capital of \$10 million, of which Congress, through the Treasury, subscribed 20 per cent, the bulk of the rest coming from foreign sources. Note issues were not to exceed the amount of capital and deposit.

(b) *Branches of the Bank.* Branches were opened in eight major commercial centres and so successful was the Bank that many former opponents on the Republican side were won over to its support. It was a great aid to the Treasury Department in handling deposits and disbursements, regulating the currency and making loans, and it was of great service to trade and industry.

(c) *Opposition to the First Bank.* The charter was not renewed in 1811 because of the following factors:

(i) The opposition of state banks, which were able to rally support against the Bank of Congress.

(ii) The feeling that the Bank was dominated by foreigners who owned 72 per cent of its stock.

(iii) A fear of centralised institutions, and a determination at state level that the federal government should operate only within the limits as set out in the constitution; some considered the First Bank of the U.S.A. to be unconstitutional.

6. The Second Bank of the U.S.A.

(a) *Composition of the Bank.* The Second Bank was a similar but larger version of the First Bank. The capital was \$35 million, of which the Treasury subscription was 20 per cent, and the President had power to nominate five of the twenty-five directors. By 1830 it had twenty-four branches operating in the main urban commercial centres. Notes could not be issued in excess of capital stock.

(b) *Threatened collapse.* The Bank came near to collapse in 1818–19. Reform of policy was effected by Langdon Cheves with the aid of a \$2 million loan from Barings, but the calling in of loans made the Bank unpopular in the West and South.

(c) *Biddle's Bank.* Nicholas Biddle became the Bank's President in 1823 and under his direction the Bank became a strong financial institution acting as the repository of the nation's specie reserves and fulfilling the role of a central bank. It is considered that the banking function was as successfully performed by the Second Bank in 1825 as by any other bank in the world at that time:

(*i*) It regulated the supply of money.

(*ii*) It restrained expansion of bank credit.

(*iii*) It protected the money market from disturbances that might result from: (1) Treasury action; (2) fluctuations in the national balance of payments; or (3) regional imbalances.

(*d*) *Unpopularity of the Bank.* Because of its restraint of bank credit and currency issues the Second Bank was unpopular in the West and South, where expansionist policies were desired. Popular opposition to the Bank was exploited by Andrew Jackson (President of the U.S.A., 1828–36) and its charter was not renewed. Refusal to prolong the life of the Bank was viewed as meaning the following:

(*i*) The ending of federal control over bank credit.

(*ii*) A removal of restriction on expansion of both credit and note issue.

(*iii*) The end of a monopolistic institution that threatened the rights of states.

(*e*) *Effects of the ending of the Second Bank.* These were as follows:

(*i*) Currency instability.

(*ii*) A loss of an equalisation force in minimising regional financial imbalance.

(*iii*) The establishment of an independent Treasury to look after government funds.

(*iv*) A shift in the centre of financial power from Chestnut Street, Philadelphia, to Wall Street, New York.

(*v*) A setback to the development of central banking in the U.S.A.

7. The Suffolk system.

(*a*) *Central bank functions.* The Suffolk Bank of Boston developed as a fairly effective type of central bank in the State of Massachusetts and by 1850 was serving as such for New England. It was not a government depository or a fiscal agent, but was able to act as a central bank because of its great prestige. Other banks kept deposits with it: country banks were compelled to do so because of threats by the Suffolk to present their banks' notes at the banks of issue for redemption at par value. It was able to operate effectively because of the great demand for bank credit which kept bank reserves at full stretch.

(b) *The Bank's operations.* The operations of the Suffolk Bank meant that banknotes in New England were redeemed at par value: fluctuations in the value of banknotes were ironed out. This was of great help to commerce and economic development.

8. The New York Safety Fund system. In 1829 the State of New York established a Safety Fund system to provide some guarantee of security for holders of banknotes and for depositors. Each bank in New York State had to contribute to the fund $\frac{1}{2}$ per cent of the bank's capital each year for six years, and each bank had to open its doors for inspection of the accounts by three commissioners every four months. The system of inspection lasted until 1843, when it was abolished, having proved somewhat ineffective. The safeguard for depositors had to be ended in 1841 because of excessive calls on the fund following the panic and closure of banks after the 1837 crisis.

Both the Suffolk system and the New York Safety Fund were somewhat unpopular because each tended to prevent misuse of bank credit and, therefore, restricted the expansion of such credit at a time when expansionary forces were pushing violently against the doors of banks. Those who wanted more rapid economic growth found displeasing any curb on credit: the ideals of sound currency and conservative growth were under constant pressure, and easy money policies were in great demand.

THE FREE BANKING ERA, 1837–63

9. Bank instability.

(a) *1837–63.* These years were ones in which American banking was unstable: large numbers of banks had to close their doors to business. There was a significant increase in the number of state banks – many of them formed under general state laws – the origin of so-called "free" banking. In New York alone some sixty banks failed between 1839 and 1860, the bulk of them failing in 1839 and 1846. The years 1837–41 and 1852–8 were the worst years for bank failures.

(b) *1834–60.* During this period the number of banks trebled, but the total capital employed failed to keep pace with the growth in numbers.

10. Free bank legislation. The "free" banking period began in 1837 when the legislature in the State of Michigan enacted a state law authorising anyone who complied with conditions set out in the Act to engage in banking. Between 1837 and 1860 thirteen other states enacted similar laws.

11. Wildcat banks. Free banking in the State of Michigan was notorious for its instability: it was here that the term "wildcat banking" originated. Bank associations were formed, currency notes were printed, state lands were purchased and then the banks set up offices in the "wilds" to make reception of notes difficult, if not impossible.

12. Currency and banks of issue.

(*a*) *Banks of issue*. In general, these banks acquired doubtful reputations, so that by 1852, in some states, such banking was prohibited for the following reasons:

    (*i*) Large numbers of counterfeit notes were in circulation.

    (*ii*) The value of notes fluctuated with the distance from the bank of issue and, often, genuine notes could be redeemed only at a substantial discount, thus creating uncertainty in the banking world.

    (*iii*) Partly because of the above, banks were seen as a cause of economic instability, encouraging undue speculation. (The Loco Foco movement, in particular, objected to banks on this score.)

    (*iv*) "Hard" money was the only generally acceptable type of currency after the closure of the Second Bank of the U.S.A.

(*b*) *Instability*. Instability varied considerably over the nation. In general banks in the East tended to be sound and those in the West to be less so.

(*c*) *Proliferation of note issues*. By 1863 there was a proliferation of note issues. Davis (*Chicago Tribune*, 1863) estimated that there were in circulation more than 14,000 different issues of banknotes, about half of which were counterfeit.

13. Summary. In 1863 American banking was in an unwholesome condition, inducing economic instability and uncertainty. No widely effective central authority in banking existed to don the mantle of stability and reliability on a bank system that was being prised apart by forces of rapid economic expansion. Rapid economic growth caused the unwise expansion of banking, which in turn, in certain areas, hindered further rapid growth.

THE NATIONAL BANK PERIOD, 1863–1913

14. Currency: the Civil War years. During the Civil War the unsatisfactory currency situation was exacerbated:

(a) Specie payments were suspended in December 1861. Gold was withdrawn from circulation.

(b) Paper currency of local banks could not be used in government transactions, and Congress, therefore, authorised the issue of United States notes – "greenbacks" – of which $150 million were issued at first, but by 1865 some $450 million were in circulation. These notes were inconvertible and depreciated considerably, in some cases by as much as 65 per cent. (It was the intention that "greenbacks" should be used during the wartime period only, but in fact they continued in use afterwards and were made convertible under the *Specie Resumption Act* of 1875.)

15. National Currency Act and National Bank Act. In 1863 a *National Currency Act* provided for a currency issue of $300 million and the *Bank Act* provided the media by which the currency was to pass into circulation.

(a) *National Bank Association.* A National Bank Association was to be set up to issue notes to the value of $900 million.

(b) *Bank Associations.* These could be formed under the Act. Five or more persons could form such an association under federal charter subject to certain minimum capital requirements:

(i) In towns with a population of more than 50,000 the minimum capital was $200,000.

(ii) In towns with a population of 6,000–50,000 the minimum capital was $100,000.

(iii) In towns with a population of less than 6,000 the minimum capital was $50,000 (subject to special Treasury approval).

(c) *Deposits.* Each association had to deposit with the Treasury of the U.S.A. registered bonds or not less than a third of the capital stock of the bank, whichever was greater, and notes to 90 per cent of the value of the bonds would be issued. (Bonds were to be valued at current market prices.)

(d) A three-tier system of banking was instituted under the Act:

(i) New York (and Chicago and St Louis in 1887) was to be a

Central Reserve City. National Banks had to have reserves of 25 per cent (of note issues and deposits) in cash.

(*ii*) Seventeen Reserve Cities (or Redemption Cities) were to have National Banks with reserves of either 25 per cent in the form of cash or 12½ per cent in cash and 12½ per cent as deposits in Central Reserve City Banks.

(*iii*) In other areas country banks were to be set up with reserves of 15 per cent completely in cash, or 6 per cent in cash and 9 per cent as deposits with Central Reserve City or Reserve City Banks.

(*e*) *Note-issue tax.* To discourage the issuing of notes by other banks the federal government, in 1865, placed a tax of 10 per cent on note issues by banks outside the National Bank system.

16. Growth of banking. The number of banks within the national system grew rapidly after the 1865 Act and the number of state banks declined until it was realised that deposit banking, and not currency issue alone, provided spheres within which banks could operate profitably. Then the number of state banks increased rapidly.

17. Advantages of the National Bank system.

(*a*) *Uniformity of currency improved.* There was a decline in note issues by other banks. National Bank notes were of general acceptability and of stable value: confidence in the paper currency was higher.

(*b*) *Government depositories.* The National Banks were convenient depositories for government funds.

(*c*) *Business confidence.* Commercial transactions were facilitated because of the confidence in the banknotes.

(*d*) *United States bonds.* The banks provided a ready market for United States bonds.

18. Weaknesses of the National Bank system.

(*a*) *Lack of stability.* The system was not really centralised: it was an aggregate of units which tended to be relatively small and not linked together sufficiently to assist each other in times of economic difficulty. Consequently, National Banks were almost as prone to failure as were state banks. Between 1893 and 1897 186 had to close their doors.

(*b*) *Concentration of reserves.* National Banks did not develop branch banking to any great extent. Reserves, therefore, were not widely spread and were confined too narrowly to one locality or one industry. (Amalgamation and branch banking would have overcome this weakness.)

(c) *Inelasticity of money supply.* Note issues were too dependent on the price and yields of United States bonds. Between 1873 and 1891 note circulation decreased from $339 million to $168 million at a time when economic growth was rapid and commerce was expanding. Some economists consider that the supply of money was too inelastic, and that the nominal supply probably lagged behind the growth in the demand for money. For this the National Bank system was to some extent responsible. The note issue needed to fluctuate more in accord with the needs of business than with the supply of United States bonds.

(d) *Treasury influence.* The National Banks were susceptible to changing demands by the Treasury in their role as national depositories. A Treasury withdrawal of funds could affect adversely the volume of currency issued.

(e) *Concentration of reserves.* Reserves tended to concentrate in New York (and later Chicago and St Louis as well), where they were lent out at short call to brokers so that interest could be earned. Security markets were affected when funds were called in by Central Reserve City Banks. There was no "lender of last resort" to replenish reserves in times of stress.

(f) *Rural difficulties.* Capital requirements for National Banks tended to be too high for rural and frontier areas: these were areas where liberal credit policies were needed to promote economic expansion. In these areas especially a flexible money supply responsive to seasonal, cyclical and secular change in the volume of transactions was needed. Account needed to be taken of the pressure on loanable funds, changes in liquidity preference and the growth in population, with rising *per capita* income, as well as the tendency for business units to increase in number and size.

19. Other features of American banking, 1863–1913.

(a) *Money supply.* Whilst silver and gold and National Bank notes increased in supply, there was a much greater increase in the use of cheques. Bank deposits increased enormously.

(b) *Investment banking.* There was a growth of investment banking, especially after 1890, with bankers securing a great measure of control in some sectors of industry. J. P. Morgan rose to a position of great influence by this means: in 1910 he and his partners had seventy-two directorships on the boards of large companies. Morgan, Baker, Stillman, etc., became active in promoting business mergers and take-overs, and in managing the New York money market: they were referred to as the "Money Trust" by the Pujo Committee of the House of Representatives in 1912. Sectors of industry particularly affected by banker influence were those in which capital requirements were large and in which the large unit had great economic advantages. Heavy industry (the United States Steel Corporation, for example), railroads, shipping, chemicals, oil, etc., were areas in which mergers could profitably be arranged.

(c) *Banker influence.* The power of bankers such as James Stillman, President of the National City Bank, was enormous. Stillman presided over a gold reserve comparable to that of the United States Treasury. His bank had deposits from more than 200 country banks and to some extent the National City Bank performed the functions of a central bank, but with little concern for other banks in times of pressure. Policies of such large banks tended to amplify rather than nullify pressure of the business cycle. In times of crisis the private interest of such a bank was paramount; obligations to its own depositors and shareholders had to be fulfilled first; there was no question of general public interest taking pride of place.

THE FEDERAL RESERVE ERA

20. The Federal Reserve Act, 1913. This Act came into effect on 30th June 1915:

(a) A *Federal Reserve Board* was set up:

(i) It had seven members, composed of the Secretary of the Treasury and the Comptroller of Currency, who were *ex-officio* members (until 1935), and five members chosen by the President of the U.S.A. From these five a Governor and Vice-Governor were elected.

(ii) The Board was to determine the policies of the Federal Reserve Banks.

(b) Twelve *Federal Reserve Banks* were formed:

(i) The country was divided into twelve Federal Reserve Areas.

(ii) Each Federal Reserve Bank was to be in a leading city of each area. City branches were allowed after the 1923 Act.

(iii) The Federal Reserve Banks were to act as central banks in their respective areas, and to act as government banks.

(iv) Old National Bank notes were replaced by Federal Reserve banknotes: no Federal Reserve Bank could pay out notes of other Federal Reserve Banks.

(v) Each Federal Reserve Bank was to have nine directors (three from the banking world, three from industry and commerce and three appointed by the Federal Reserve Board.)

(c) All *National Banks* were invited to join the Federal Reserve system or surrender their charters: other banks were invited to join as well. Member banks had the same designation as under the National Bank system – Central Reserve, City, Reserve City or Country Banks according to the population of the cities and towns concerned. Reserve requirements demanded varied with the size of the town or city.

21. The Federal Reserve system.

(a) *Federal Reserve Act*, 1913. This was passed to remedy the deficiencies of the National Bank system but, in fact, American banking continued to be relatively unstable in the inter-war period until the *Emergency Banking Act* of 1933 restored some stability in the 1930s. Under the *Federal Reserve Act* the country was divided into twelve areas each with a Federal Reserve Bank which acts as a central bank controlling its own note issue.

(b) *Policies of Federal Reserve Banks*. Policies are co-ordinated by a Federal Reserve Board. All National Banks became members of the Federal Reserve system together with state banks which met stipulated requirements and chose to join.

(c) *Functions*. Federal Reserve Banks were set up to provide financial services for the federal government, to issue currency according to commercial needs and to maintain credit conditions essential to healthy economic activity by varying deposit require-ments, the use of open market operations, varying re-discount rates and the use of moral persuasion.

22. Dualism of authority.
During the First World War the Treasury dictated policy because of war-time demands on the supply of money, but in 1919 the Federal Reserve Board regained control of currency issues. For a time in the 1920s dualism of authority developed because of the strength of the New York Federal Reserve Bank, which became important in international relations. The *McFadden Act* of 1923 permitted member banks to open branches, and there was a trend towards amalgamation, but, even so, failures were high: estimates suggest that between 1921 and 1929 close on 1,000 member banks failed. Some 30,000 banks existed altogether, many of them small, with inadequate

capital and too closely identified with particular localities. Nearly 6,000 state non-member banks failed between 1921 and 1929. There was an unwise expansion of loans against securities and of investment banking: consequently the 1929–33 depression had disastrous effects on banks. Some 5,099 failed in 1930–2 – the hearts of the financial world began to flutter and in too many cases to cease beating at all, necessitating federal intervention to prevent a complete collapse of banking.

23. The Emergency Banking Act, 1933. In 1933 the *Emergency Banking Act* was passed, widening the powers of the Federal Reserve Banks and setting up a Reconstruction Finance Corporation to supply capital to banks. A Federal Deposit Insurance Corporation was set up to guarantee small bank deposits. Banking slowly recovered from the 1929 crash.

24. The Banking Act, 1935. The *Banking Act* of 1935 gave more centralised control over market operations, discount rates and reserve requirements of banks. State banks with deposits of $1 million or more were required to join the Federal Reserve system by 1942 if they wished to qualify for federal insurance of deposits. By 1960 there were 13,000 banks in active existence: twenty-five had assets of more than $1,000 billion each, and 10,000 had less than $1 million each. About 50 per cent of all banks were in the Federal Reserve system; they held 85 per cent of all bank assets, and the ten largest banks possessed 20 per cent of all bank resources. State banks in 1960 were still confined to banking within the state borders but banking co-operation was much closer than previously. Banking was on a much sounder footing, with the Federal Reserve Board exercising a much closer supervision over banks than before the Second World War.

MONEY SUPPLY

25. 1792–1834.

(*a*) The currency was bimetallic but only silver was in circulation as hard currency. The silver dollar was the basic coinage.

(*b*) From 1791 to 1811 the notes of the First Bank of the U.S.A. were in circulation and from 1816 to 1836 those of the Second Bank were circulated.

(*c*) In addition the notes of the various state banks circulated.

(*d*) Bills of exchange were utilised as money.

26. 1834–63.

(a) In 1834 the gold–silver ratio was revised, and with gold and silver dollars in use the currency was more truly bimetallic.

(b) This was the era of "free" banking with a proliferation of note issues by state banks.

(c) Bills of exchange continued to be used as a form of money.

27. 1863–1913.

(a) Treasury notes (greenbacks) were issued during the Civil War to the extent of $450 million. In 1875 they were made convertible, and in 1879 were redeemable at par value, but the amount in circulation was reduced to $347 million.

(b) State bank notes continued to be issued, but the tax imposed on them caused many issues to be replaced by National Bank notes.

(c) There was considerable controversy about the role of silver in the monetary system. In 1873 the silver dollar was omitted from the list of authorised coins, but in 1878 the *Bland–Allison Act* permitted the Treasury to buy $2 million to $4 million of silver each month for coinage into dollars, and silver certificates could be issued in denominations of $10 and more. The *Sherman Silver Purchase Act* of 1890 provided for larger silver purchases, but it was repealed in 1893.

(d) The gold standard was adopted in 1900, although silver dollars were made legal tender.

28. 1913–50.

(a) Federal Reserve Bank notes were issued upon the deposit of government bonds, and from 1933 upon the deposit of commercial paper.

(b) The gold standard was virtually abandoned in 1933 and great efforts were made to expand the money supply in the hope that it would help pull the nation out of the depression.

PROGRESS TEST 19

1. Discuss the nature of the First Bank of the U.S.A. (5)
2. Why was the Second Bank of the U.S.A. unpopular? (6)
3. Discuss the "free" banking era in the U.S.A. in the period 1836 to 1863. (6, 9–13)
4. How successful was the *National Bank Act* of 1863? (14–19)
5. Explain the operation of the Federal Reserve system. (20–24)
6. Why were *Bank Acts* necessary in the 1930s? (22–24)

APPENDIX I

BIBLIOGRAPHY

* indicates books and articles for further reference.
† indicates essential reading.

Abbreviation of journals

E. H. R. Economic History Review
E. J. Economic Journal
J. E. H. Journal of Economic History
O. E. P. Oxford Economic Papers

PART ONE: INTRODUCTION

Books

BRITAIN

†Ashworth, W.: *An Economic History of England, 1870–1939* (Methuen, 1960).

*Bruce, M.: *The Coming of the Welfare State*, fourth edition (Batsford, 1968).

*Deane, P., and Cole, W. A.: *British Economic Growth, 1688–1959*, second edition (Cambridge University Press, 1967).

†Hobsbawm, E. J.: *Industry and Empire* (Pelican, 1969).

†Pollard, S.: *The Development of the British Economy, 1914–67* (Arnold, 1968).

†Sayers, R. S.: *A History of Economic Change in England, 1880–1939* (Oxford Paperbacks, 1967).

U.S.A.

*Hacker, L. M.: *American Capitalism* (Anvil, 1957).

†Nash, G. D.: *Issues in American Economic History* (Heath, 1964).

†North, D. C.: *Growth and Welfare in the American Past* (Prentice-Hall, 1966).

*Robertson, R. M.: *History of the American Economy* (Harcourt, Brace & World, 1955).

†Williamson, H. F. (ed.): *Growth of the American Economy* (Prentice-Hall, 1958).

Articles

*Aldcroft, D. H.: "Economic Growth in Britain in the Inter-war Years."
  E.H.R., Second Series, vol. 20.
*Beales, H. L.: "The Great Depression in Industry and Trade." *E.H.R.*,
  First Series, vol. 5.
†Kemmerer, D. L.: "The Changing Pattern of American Economic
  Development." *J.E.H.*, vol. 16.
†Musson, A. E.: "The Great Depression in Britain, 1873–96: A Re-
  appraisal." *J.E.H.*, vol. 19.
*Opie, R., "Anglo-American Economic Relations in Wartime." *O.E.P.*,
  vol. 9.
†Robinson, E. A. G.: "The Changing Structure of the British Economy."
  E.J., vol. 64.
†Rostow, W. W.: "The Stages of Economic Growth." *E.H.R.*, Second
  Series, vol. 12.
†Rostow, W. W.: "The Take-off into Self-sustained Growth." *E.J.*,
  vol. 66.
*Webber, B., and Handfield Jones, S. J.: "Variations in the Rate of
  Economic Growth in the U.S.A., 1869–1939." *O.E.P.*, vol. 6.
†Williamson, J. G.: "The Long Swing: Comparisons and Interactions
  Between British and American Balance of Payments, 1820–1913."
  J.E.H., vol. 22.

Part Two: POPULATION

Books

BRITAIN

*Cairncross, A. K.: *Home and Foreign Investment, 1870–1913* (Cambridge
  University Press, 1953).
*Deane, P., and Cole, W. A.: *British Economic Growth, 1688–1959*,
  second edition (Cambridge University Press, 1967).
†Drake, M.,: *Population in Industrialisation* (Methuen, 1969).
†Glass, D. V., and Eversley, D. C.: *Population in History* (Arnold, 1965).
†Hubback, E.: *The Population of Britain* (Penguin Books, 1947).
*Mitchell, B., and Deane, P.: *Abstract of British Historical Statistics*
  (Cambridge University Press, 1962).
†Pollard, S.: *The Development of the British Economy, 1914–67* (Arnold,
  1968).
†Sayers, R. S.: *A History of Economic Change in England, 1880–1939*
  (Oxford Paperbacks, 1967).
†Wrigley, E. A.: *Population and History* (Weidenfeld & Nicolson, 1969).
†*The Report of the Royal Commission on Population, 1949* (H.M.S.O.).

U.S.A.

*Berthoff, R. T.: *British Immigrants in Industrial America, 1790–1950*
  (Russell & Russell, 1953).

†Faulkner, H. U.: *American Economic History*, eighth edition (Harper & Row, 1960).

*Hacker, L. M.: *Major Documents in American Economic History* (Anvil, 1961), Vol. 2.

†Harris, S. (ed.): *American Economic History* (McGraw-Hill, 1961).

†Jones, M.: *American Immigration* (University of Chicago, 1960).

*Lebergott, S.: *Manpower in Economic Growth: The U.S.A. Record Since 1860* (McGraw-Hill, 1964).

†North, D. C.: *Growth and Welfare in the American Past* (Prentice-Hall, 1966).

†Scott, F. D.: *Emigration and Immigration* (Macmillan (New York), 1963). (An American Historical Association pamphlet.)

*Thomas, B.: *Migration and Economic Growth* (Cambridge University Press, 1954).

Articles

†Conrad, A. H., *et al.*: "Slavery as an Obstacle to Economic Growth in the United States." *J.E.H.*, vol. 27.

†Danhof, C. H.: "Economic Validity of the Safety-valve Doctrine." *J.E.H.*, vol. 1.

*Dovring, F.: "European Reaction to the Homestead Act." *J.E.H.*, vol. 22.

†Marshall, T. H.: "The Population of England and Wales from the Industrial Revolution to the World War." *E.H.R.*, First Series, vol. 5.

*Meade, J. E.: "Population Explosion, the Standard of Living and Social Conflict." *E.J.*, vol. 77.

*Shepperson, W. S.: "Industrial Emigration in Early Victorian Britain." *J.E.H.*, vol. 13.

PART THREE: AGRICULTURE

Books

BRITAIN

†Ashworth, W.: *An Economic History of England, 1870–1939* (Methuen, 1960), Chapter 3.

†Chambers, J. D., and Mingay, G. E.: *The Agricultural Revolution, 1750–1880* (Batsford, 1966).

†Court, W. H. B.: *British Economic History, 1870–1914* (Cambridge University Press, 1965), Chapter 2.

*Deane, P., and Cole, W. A.: *British Economic Growth, 1688–1959*, second edition (Cambridge University Press, 1967).

*Ernle, Lord: *English Farming, Past and Present* (Heinemann, 1961 edition, with an introduction by O. R. McGregor and G. E. Fussell).

†Jones, E. L.: *The Development of English Agriculture, 1815–1873* (Macmillan, 1968).

†Jones, G. P., and Pool, A. G.: *A Hundred Years of Economic Development in Britain, 1840–1940*. (Duckworth & Co., 1940), Chapters 3, 10 and 16.

†Orwin, C. S., and Whetham, E. H.: *History of British Agriculture, 1846–1914* (Longmans 1964).

*Pollard, S.: *The Development of the British Economy, 1914–67* (Arnold, 1968).

†Thompson, F. M. L.: *English Landed Society in the Nineteenth Century* (Routledge & Kegan Paul, 1963).

U.S.A.

*Billington, R. A.: *The Westward Movement in the United States* (Anvil, 1959).

*Faulkner, H. U.: *American Economic History*, eighth edition (Harper & Row, 1960).

†Faulkner, H. U.: *The Decline of Laissez-faire* (Holt, Rinehart & Winston, 1961).

†Hacker, L. M.: *Major Documents in American Economic History* (Anvil, 1961), vols. 1 and 2.

†Mitchell, B.: *Depression Decade* (Holt, Rinehart & Winston, 1962).

*North, D. C.: *Growth and Welfare in the American Past* (Prentice-Hall, 1966).

*Shannon, F. A.: *American Farmers' Movements* (Anvil, 1957).

†Shannon, F. A.: *The Farmers' Last Frontier* (Holt, Rinehart & Winston, 1961).

†Soule, G.: *Prosperity Decade* (Holt, Rinehart & Winston, 1962).

†Williamson, H. F. (ed.): *Growth of the American Economy* (Prentice-Hall, 1958).

Articles

*Collins, E. J. T.: "Harvest Technology and Labour Supply in Britain, 1790–1870." *E.H.R.*, Second Series, vol. 22.

†Fairlie, S.: "The Corn Laws and British Wheat Production, 1829–76." *E.H.R.*, Second Series, vol. 22.

†Fletcher, T. W.: "The Great Depression of English Agriculture, 1873–96." *E.H.R.*, Second Series, vol. 13.

*Hunt, E. H.: "Labour Productivity in English Agriculture, 1850–1914." *E.H.R.*, Second Series, vol. 20.

*Jones, E. L.: "The Agricultural Labour Market in England, 1793–1872. *E.H.R.*, Second Series, vol. 17.

*Moore, D. C.: "The Corn Laws and High Farming." *E.H.R.*, Second Series, vol. 18.

†Rasmussen, W. D.: "The Impact of Technological Change on American Agriculture, 1862–1962." *J.E.H.*, vol. 22.

*Saloutos, T.: "Land Policy and its Relation to Agricultural Production and Distribution, 1862–1933." *J.E.H.*, vol. 22.

*Saloutos, T.: "The Spring Wheat Farming in a Maturing Economy, 1870–1920." *J.E.H.*, vol. 6.
†Thompson, F. M. L.: "The Land Market in the Nineteenth Century." *O.E.P.*, vol. 9.

PART FOUR: INDUSTRIAL DEVELOPMENT

For those studying Britain, there are informative chapters in the books by W. Ashworth, W. H. B. Court, P. Deane and W. A. Cole, E. J. Hobsbawm, S. Pollard, and R. S. Sayers referred to for Parts One to Three. For the U.S.A. reference should be made to the chapters on industry in the books by H. U. Faulkner, S. Harris (ed.), B. Mitchell, G. D. Nash (ed.), D. C. North, R. M. Robertson, G. Soule and H. F. Williamson (ed.), all of which are also referred to for Parts One to Three.

In addition reference should be made to the works listed below.

Books

BRITAIN

*Carus-Wilson, E. M. (ed.): *Essays in Economic History* (Arnold, 1963), vol. 1, pp. 314–415.
†Checkland, S. G.: *The Rise of Industrial Society in England, 1815–1885.* Longmans, 1964.
*Clapham, J. H.: *An Economic History of Modern Britain* (Cambridge University Press), vols. 2 and 3. (These are massive, detailed works which were published in 1932 and 1938 respectively, and have since been reprinted several times. The latest edition should be obtained.)
†Mowat, C. L.: *Britain Between the Wars, 1918–1940* (Methuen, 1955). (The standard work on this period.)
†Saul, S. B.: *The Myth of the Great Depression* (Macmillan, 1969).
†Youngson, A.: *The British Economy, 1920–57* (Allen & Unwin, 1960).

U.S.A.

*Cochran, T. C.: *Basic History of American Business* (Anvil, 1959).
†Deglar, C. N.: *The Age of the Economic Revolution, 1876–1900* (Scott, Foresman & Co., 1967).
*Habakkuk, H. J.: *British and American Technology in the Nineteenth Century* (Cambridge University Press, 1962).
*Hacker, L. M.: *The Triumph of American Capitalism* (Columbia University Press, 1940).
*Hexner, E. P., and Walters, A.: *International Cartels* (Chapel Hill, 1946).
*Jones, P. d'A.: *The Consumer Society* (Penguin Books, 1965).
†Kirkland, E. C.: *Industry Comes of Age* (Holt, Rinehart & Winston, 1961).

Articles

†Aldcroft, D. H.: "Economic Growth in Britain in the Inter-war Years, A Reassessment." *E.H.R.*, Second Series, vol. 20.

†Aldcroft, D. H.: "The Entrepreneur and the British Economy, 1870–1914." *E.H.R.*, Second Series, vol. 17.

†Ames, E. A.: "Trends, Cycles and Stagnation in United States Manufacturing Since 1860." *O.E.P.*, vol. 11.

†Coppock, D. J.: "British Industrial Growth, 1873–1896." *E.H.R.*, Second Series, vol. 17.

*Frost, R.: "The Macmillan Gap, 1931–53." *O.E.P.*, vol. 6.

*Habakkuk, H. J.: "Fluctuations in House Building in Britain and the United States in the Nineteenth Century." *J.E.H.*, vol. 22.

*Hoffman, C.: "The Depression of the Nineties." *J.E.H.*, vol. 16.

†Musson, A. E.: "British Industrial Growth, 1873–1896: A Balanced View." *E.H.R.*, Second Series, vol. 17.

*Payne, P. L.: "The Emergence of the Large-scale Company in Britain, 1870–1914." *E.H.R.*, Second Series, vol. 20.

†Richardson, H. W.: "The New Industries Between the Wars." *O.E.P.*, vol. 13.

*Richardson, H. W.: "Over-commitment in Britain Before 1930." *O.E.P.*, vol. 17.

†Rosenberg, N.: "Technological Change in the Machine Tool Industry, 1840–1910." *J.E.H.*, vol. 23.

†Saul, S. B.: "The Market and the Development of the Mechanical Engineering Industries in Britain, 1860–1914." *E.H.R.*, Second Series, vol. 20.

PART FIVE: TRANSPORT

Books

BRITAIN

*Checkland, S. G.: *The Rise of Industrial Society in England, 1815–85* (Longmans, 1964).

†Kellett, J. R.: *The Impact of Railways on Victorian Cities.* (Routledge & Kegan Paul, 1969).

†Pollard, S.: *The Development of the British Economy, 1914–67* (Arnold, 1968).

†Savage, C.: *An Economic History of Transport* (Hutchinson University Library, 1959).

*Sturmey, S. C.: *British Shipping and World Competition* (Athlone Press, 1962).

†Thornton, R. H.: *British Shipping* (Cambridge University Press, 1959).

U.S.A.

*Degler, C.: *The Age of the Economic Revolution* (Scott, Foresman & Co., 1967).

*Fogel, R. W.: *Railroads and American Economic Growth* (John Hopkins Press, 1964).

†Kirkland, E. C.: *Men, Cities and Transportation* (Harvard University Press, 1948).

†Nash, G. D.: *Issues in American Economic History* (Heath, 1964), Chapter 18.

†Taylor, G. D.: *The Transportation Revolution* (Holt, Rinehart & Winston, 1951).

†Williamson, H. F.: *Growth of the American Economy* (Prentice-Hall, 1958), Chapters 7, 19 and 35.

Articles

†Cootner, P. H.: "The Role of the Railroads in United States Economic Growth." *J.E.H.*, vol. 23.

†David, P. A.: "Transport Innovation and Economic Growth." *E.H.R.*, Second Series, vol. 22.

†Fogel, R. W.: "Railroads as an Analogy to the Space Effort," *E.J.*, vol. 76.

†Hawke, G. R., and Reed, M. C.: "Railway Capital in the United Kingdom in the Nineteenth Century." *E.H.R.*, Second Series, vol. 22.

†Jenks, L. H.: "Britain and American Railway Development." *J.E.H.*, vol. 11.

†Jenks, L. H.: "Railroads as an Economic Force in American Development." *Essays in Economic History*, edited by Carus Wilson (Arnold, 1964), vol. 3.

*Leonard, W. N.: "The Decline of Railroad Consolidation." *J.E.H.*, vol. 9.

†McClelland, P. D.: "Railroads, American Growth and the New Economic History: A Critique." *J.E.H.*, vol. 28.

†Mitchell, B. R.: "The Coming of the Railway and United Kingdom Economic Growth." *J.E.H.*, vol. 24.

*North, D. C.: "Ocean Freight Rates and Economic Development, 1750–1913." *J.E.H.*, vol. 18.

*Pollard, S.: "British and World Shipbuilding, 1890–1914." *J.E.H.*, vol. 17.

PART SIX: LABOUR MOVEMENTS

Books

BRITAIN

*Citrine, N. A.: *Citrine's Trade Union Law* (Stevens, 1967).

†Cole, G. D. H.: *A Short History of the British Working Class Movement* (Allen & Unwin, 1948).

*Court, W. H. B.: *British Economic History, 1870–1914* (Cambridge University Press, 1965), Chap. 7.

*Hobsbawm, E. J.: *Labouring Men* (Weidenfeld & Nicolson, 1965).

†Pelling, H.: *A History of British Trade Unionism* (Penguin Books, 1965).

†Phelps-Brown, E. H.: *The Growth of British Industrial Relation* (Macmillan, 1959).

†Pollard, S.: *The Development of the British Economy, 1914–1967* (Arnold, 1968).

U.S.A.

†Dulles, F. R.: *Labor in America* (Crowell, 1964).
†Harris, S. (ed.): *American Economic History* (McGraw-Hill, 1961).
†Pelling, H.: *American Labor* (Chicago University Press, 1960).
†Taft, P.: *Organised Labor in American History* (Harper & Row, 1964).

Articles

†Cole, G. D. H.: "British Trade Unions in the Third Quarter of the Nineteenth Century." *Essays in Economic History,* edited by Carus-Wilson (Arnold, 1964), vol. 3.
*Grob, G. N.: "Reform Unionism: The National Labor Union." *J.E.H.,* vol. 14.
*Murphy, C. G. S., and Zellner, A.: "Sequential Growth: The Labor Safety Valve Doctrine and the Development of American Unionism." *J.E.H.,* vol. 19.
†Sires, R. V.: "Labor Unrest in England, 1910–1914." *J.E.H.,* vol. 15.

PART SEVEN: FOREIGN TRADE

Books

BRITAIN

†Ashworth, W.: *An Economic History of England, 1870–1939* (Methuen, 1960), Part 1, Chap. 6, and Parts 2 and 3.
†Court, W. H. B.: *British Economic History, 1870–1914* (Cambridge University Press, 1965), Chap. 4.
†Jones, G. P., and Pool, A. G.: *A Hundred Years of Economic Development* (Duckworth, 1940), Chaps. 9, 15 and 18.
*Khan, A. E.: *Great Britain in the World Economy* (Columbia U.P., 1946).
†Pollard, S.: *The Development of the British Economy, 1914–1967* (Arnold, 1968).
†Saul, S. B.: *Studies in British Overseas Trade* (Liverpool University Press, 1960).
†Sayers, R. S.: *A History of Economic Change in England, 1880–1939* (Oxford University Press, 1967).
*Yates, P. L.: *Forty Years of Foreign Trade* (Allen & Unwin, 1959).

U.S.A.

The nine volumes of *Economic History of the United States,* published by Holt, Rinehart & Winston, have chapters on aspects of foreign trade. Especially useful are these three volumes:

   Faulkner, H. U.: *The Decline of Laissez-faire* (1951).
   Mitchell, B.: *Depression Decade* (1947).
   Soule, G.: *Prosperity Decade* (1947).
†Ashworth, W.: *A Short History of the International Economy, 1850–1960* (Longmans, 1962 edition).

†Faulkner, H. U.: *American Economic History*, eighth edition (Harper & Row, 1959), Chaps. 25, 26, 29–32.
†Jones, P. d'A.: *The Consumer Society* (Penguin Books, 1965), Chaps. 3, 5, 6, 8–10).
*Jones, P. d'A.: *An Economic History of the United States Since 1783* (Routledge & Kegan Paul, 1956), Chaps. 5, 9. 13 and 14.
*Robertson, R. M.: *History of the American Economy* (Harcourt, Brace & World), 1955.
†Williamson, H. F. (ed.): *Growth of the American Economy* (Prentice-Hall, 1958), Chaps. 12, 27 and 41.

Articles

†Ford, A. G.: "Bank Rate, the British Balance of Payments and the Burdens of Adjustment, 1870–1914." *O.E.P.*, vol. 16.
*Gallagher, T.: "The Imperialism of Free Trade." *E.H.R.*, Second Series, vol. 7.
†Imlah, A. H.: "The Terms of Trade of the United Kingdom, 1789–1913." *J.E.H.*, vol. 10.
†Kindleberger, C. P.: "Foreign Trade and Economic Growth, 1850–1913." *E.H.R.*, Second Series, vol. 14.
*Opie, R.: "Anglo-American Economic Relations in Wartime." *O.E.P.*, vol. 9.
†Robinson, E. A. G.: "The Changing Structure of the British Economy." *E.J.*, vol. 64.
†Rostow, W. W.: "The Terms of Trade in Theory and Practice." *E.H.R.*, Second Series, vol. 3.
†Saul, S. B.: "Britain and World Trade, 1870–1914." *E.H.R.*, Second Series, vol. 7.
†Simon, M. L., and Novack, D. E.: "Some Dimensions of the American Commercial Invasion of Europe, 1871–1914." *J.E.H.*, vol. 24.
†Thistlethwaite, F.: "Atlantic Partnership." *E.H.R.*, Second Series, vol. 7.
†Williamson, J. G.: "The Long Swing: Comparisons and Interactions between British and American Balance of Payments, 1820–1913." *J.E.H.*, vol. 22.

Part Eight: BANKING

Books

BRITAIN

†Cameron, R., *et al.*: *Banking in the Early Stages of Industrialisation* (Oxford University Press, 1967).
*Clapham, J. H.: *The Bank of England* (Cambridge University Press, 1944), vols. 1 and 2.
†Crick, W. F., and Wadsworth, J. E.: *A Hundred Years of Joint Stock Banking* (Hodder and Stoughton, 1936).

*Gregory, T. E.: *Select Statutes, Documents and Reports Relating to British Banks, 1832–1928* (Oxford University Press, 1929).

*Morgan, E. V.: *Central Banking in Theory and Practice* (Cass, 1965).

†Pressnell, L. S.: *Country Banking in the Industrial Revolution* (Clarendon Press, Oxford, 1956).

*Sayers, R. S.: *Lloyds Bank in the History of English Banking* (Clarendon Press, Oxford, 1957).

U.S.A.

†Carson, Dean (ed.): *Banking and Monetary Studies* (Irwin, 1963).

†Hammond, Bray: *Banks and Politics in America from the Revolution to the Civil War* (Princeton University Press, 1957).

*Redlich, F.: *The Molding of American Banking* (Johnson Reprint, 1968)

†Trescott, P. B.: *Financing American Enterprises: The Story of Commercial Banking* (Harper & Row, 1963).

†Williamson, H. F. (ed.): *Growth of the American Economy* (Prentice-Hall, 1958), Chaps. 13, 16, 28, 29, 42 and 43.

Articles

†Checkland, S. G.: "The Mind of the City, 1870–1914." *O.E.P.*, vol. 9.

*Clayton, G., and Osborn, W. T.: "Insurance Companies and the Finance of Industry." *O.E.P.*, vol. 10.

†Ford, A. G.: "Notes on the Working of the Gold Standard Before 1914." *O.E.P.*, vol. 12.

†Frost, R.: "The Macmillan Gap, 1931–53." *O.E.P.*, vol. 6.

*Gaskin, M.: "Anglo-Scottish Banking Conflicts, 1874–1881." *E.H.R.*, Second Series, vol. 12.

*King, W.: "The Bank of England." *E.H.R.*, Second Series, vol. 15.

*Segal, H. H., and Simon, M.: "British Foreign Capital Issues, 1865–1894." *J.E.H.*, vol. 21.

EXAMINATION TECHNIQUE

1. Read the paper carefully. The candidate should read the question paper, selecting questions in order of preference. He should make sure that the *correct* number of questions is selected, and take care to note whether the paper is divided into parts with a consequent limitation on the choice of questions.

2. Allocate enough time for each question. It is important to allocate the time allowed for the examination so that approximately the same length of time is spent over each question selected to be answered. If time is short for the last answer a summary in note form covering all the essential points is more satisfactory than producing an incomplete answer. It is essential to answer as fully as possible *all* the required questions, but, obviously, there is a tendency to answer more fully in the first, second and third answers than in the fourth, or, when required, the fifth one. The candidate should try to reveal a complete grasp of the subject in at least three answers.

3. Read each question carefully. Each question selected to be answered should be read with extreme care so that the crux, or essential point, of the question is clearly understood, and the answer should be logical and relevant. One should avoid producing answers that veer off from the main points of the questions.

4. Plan your answer. To produce a good answer in economic or social history requires of the candidate skills not unlike those used by a good lawyer arguing a case in a court of law. The essential points should be made, supported by facts and evidence from documents and the works of authoritative writers, concluding with a summary of the evidence on which assertions might be based. This can be done more easily if a plan of the answer is sketched out initially.

5. Refer to authorities. It should be borne in mind that frequently there is no conclusive answer that can be given. The answer should reveal knowledge of all facets of the question and should reveal one's breadth and depth of reading by reference to the major specialist works (books and articles) on the subject.

6. Take account of economic theory. There should be a relating of economic theory to the history: answers should take the form of economic analysis when possible.

7. Take account of all relevant factors. No one sector of the economy such as banking, transport, etc., should be considered in isolation. An awareness of the impact of one sector on other sectors should be revealed, and, obviously, there are linkages with the social and political spheres that might require mentioning. However, irrelevant digressions should be avoided.

TEST PAPERS

THREE hours should be alloted for each test paper, from which four questions should be answered. Remember that clarity of expression is very important. (Except where indicated otherwise the questions are of the type to be expected at G.C.E. "A" Level examinations.)

Abbreviations

B.Sc.(Econ.) London University, B.Sc.(Econ.).

S.C.E. Scottish Certificate of Education, Higher Grade.

Test Paper 1

1. Discuss the influence of immigration upon the economic development of the U.S.A. between 1850 and 1914. (*B.Sc.(Econ.), Part I*)

2. For what reasons was the *Bank Charter Act* of 1844 passed? How successful was it? (*S.C.E., Paper I*)

3. To what extent did Britain's farmers suffer economic difficulties after 1875 because of the expansion of American agriculture?

4. Discuss the major economic effects of the American Civil War. (*B.Sc.(Econ.), Part I*)

5. Why did birth-rates and death rates in Britain fall after 1875? (*B.Sc.(Econ.), Part I*)

6. What social and economic factors lay behind the increased industrial unrest in Britain in the period 1906–1914? (*B.Sc.(Econ.), Part II*)

7. Why were the "old" industries in such a bad way between 1921 and 1938? (*S.C.E., Paper I*)

8. Discuss the development of American foreign trade between 1900 and 1939. (*B.Sc.(Econ.), Part I*)

9. What seems to you to be the causes and the long-term significance of the Wall Street crash of 1929? (*S.C.E., Paper II*)

10. Examine the recovery of the British economy after 1933. (*B.Sc.(Econ.), Part II*)

11. How far was government policy effective in helping the United States to move out of the depression of the 1930s? (*S.C.E., Paper II*)

12. What were the main reasons for the increase in governmental intervention in economic affairs in Britain and the U.S.A. in the years 1900–1939? (*B.Sc.(Econ.), Part I*)

Test Paper 2

1. Compare the effects of railway building upon the American and British economies. (*B.Sc.(Econ.), Part I*)

2. Examine the significance of foreign trade in the economic life of Britain and the U.S.A. respectively between 1865 and 1914. (*B.Sc. (Econ.), Part I*)

3. By what steps and for what reasons was Free Trade abandoned in the twentieth century? (*S.C.E., Paper I*)

4. Why were there so many demands for monetary reform in the United States between 1865 and 1900? (*S.C.E., Paper II*)

5. Analyse changes in the structure of British banking between 1850 and 1921.

6. How successful was the American *National Bank Act* of 1864? (*B.Sc.(Econ.), Part II*)

7. To what extent did governments intervene in economic affairs in the U.S.A. and Britain before 1914?

8. Analyse changes in America's foreign trade between 1900 and 1939.

9. Examine the main changes in the structure of English industry between 1900 and 1939. (*B.Sc.(Econ.), Part I*)

10. Estimate the economic and social significance of the Model "T" Ford car. (*S.C.E., Part II*)

11. Assess the reasons for the weaknesses of American banking in the inter-war years.

12. How effective was the New Deal in solving American economic problems during the 1930s? (*B.Sc.(Econ.), Part I*)

Test Paper 3

1. Examine the special problems of British agriculture after 1873 and discuss the response to them. (*S.C.E , Part I*)

2. What were the economic effects of the American Civil War?

3. Why did it take till the 1880s for steam to replace sail? (*S.C.E., Part I*)

4. Give reasons for the anti-trust legislation in the United States after 1880. (*S.C.E., Part II*)

5. Compare and contrast trade-union development in England and the U.S.A. between 1870 and 1939. (*B.Sc.(Econ.), Part I*)

6. Examine *either* the causes *or* the consequences of the growth in the size of business units in America between 1865 and 1914. (*B.Sc.(Econ.), Part I*)

7. How do you explain the persistent fall in English food prices between 1870 and 1900? (*B.Sc.(Econ.), Part I*)

8. Discuss the development of American foreign trade between 1900 and 1950.

9. To what extent can railroads be considered analagous to the space effort?

10. What were the main features of the American boom of the 1920s? (*B.Sc.(Econ.), Part I*)

11. Discuss the problems of *either* the British shipbuilding industry *or* the coal industry between the wars.

12. Compare and contrast the ways in which Britain and the U.S.A. attempted to recover from economic depression in the 1930s.

Test Paper 4

1. What were the effects of the "moving **frontier**" on American economic life between 1850 and 1914? (*B.Sc.(Econ.), Part I*)

2. Discuss the consequences for the British economy of the system of multilateral trade that developed between 1870 and 1914. (*B.Sc.(Econ.), Part II*)

3. Outline and account for the main developments in American trade-unionism between 1860 and 1917. (*S.C.E., Paper II*)

4. In what ways did railways influence the development of the U.S.A. in the second half of the nineteenth century? (*S.C.E., Paper II*)

5. Discuss the development of American banking between 1863 and 1914.

6. What factors caused changes in the location of British industry between 1918 and 1939? (*B.Sc.(Econ.), Part I*)

7. Compare the increase in the size of business units in Britain and America between 1900 and 1950.

8. Indicate and explain the main features in British trade union history between 1900 and 1926. (*B.Sc.(Econ.), Part I*)

9. In what ways was the structure of British agriculture changed between 1850 and 1950?

10. Explain the difficulties experienced by both British and U.S.A. railway companies after 1920.

11. Discuss the view that the U.S.A. economic expansion of the 1920s was finally checked by the insufficiency of consumer demand. (*B.Sc. (Econ.), Part II*)

12. Account for fluctuations in the volume of American foreign trade between 1900 and 1950.

INDEX